ready, set, green

ready, set, green

Eight Weeks to Modern Eco-Living

from the Experts

at TreeHugger.com

Graham Hill and
Meaghan O'Neill

Ⓥ VILLARD NEW YORK

A Villard Books Trade Paperback Original

Published in the United States by Villard, an imprint of The Random House Publishing Group, a division of Random House, Inc., New York.

VILLARD and "V" CIRCLED Design are registered trademarks of Random House, Inc.

LIBRARY OF CONGRESS CATALOGING-IN-PUBLICATION DATA

Hill, Graham.
Ready, set, green : eight weeks to modern eco-living / Graham Hill and Meaghan O'Neill.
p.   cm.
Includes bibliographical references.
ISBN: 978-0-345-50308-4 (pbk.)
1. Sustainable living.   2. Energy conservation.
3. Green products.   4. Self-reliant living.
I. O'Neill, Meaghan. II. Title.
GF78.H55 2008
640—dc22   2007043724

Printed in the United States of America
on 100% post-consumer recycled stock

www.atrandom.com

9 8 7 6 5 4 3 2 1

First Edition

Book design by Liz Cosgrove

For Nicholas,
and all the littlest TreeHuggers

contents

ready, set, green

1

The Future Is Green

Look out your window. What do you see? A paved street and electrical wires? Meadows and birds? A farm full of cows? Whatever surrounds you, that's the environment. And whether it was created by Mother Nature or the municipal works department, humans aren't separate from it. Just as hurricanes, floods, and tornadoes have an effect on our well-being, we have an effect on nature, polluting water via our factories and homes, reducing mountains to piles of coal that we burn for energy, packing landfills with our used-up cars and electronics packaging. Luckily, it turns out we also have the power to clean up after ourselves.

At TreeHugger.com, the website dedicated to modern green living, we believe that cutting-edge ideas, technology, and design—and, more important, people with the right attitude—can help save the environment. This book was conceived to help readers develop an understanding of existing eco dilemmas, and to empower them to help reverse the problems. We don't have all the answers; no one does. But we believe that individuals do have the power to "green" the planet. Your dollars count. Your vote counts. Your actions count. And when millions of people do the right thing, it can have a serious impact.

A BRIEF HISTORY OF ENVIRONMENTALISM

In the mid-eighteenth century, the industrial revolution changed life as humans knew it. Local economies that produced and sold goods made primarily from biodegradable parts gave way to economies of mass-produced items that could be shipped all over the world. It was a time of great achievement and hope, but also of great innocence and ignorance—when people could not fathom that natural resources could someday become scarce or even dry up altogether.

By the late 1800s and early 1900s, the need for land conservation became apparent to people such as John Muir and President Theodore Roosevelt, the latter of whom set aside more land for national parks and nature preserves—194 million acres by 1909—than all his predecessors combined. It wasn't until the 1960s, however, that the modern environmental movement was born. Utopian idealists dreamed of living off the land and sticking it to the man. Their goals were lofty, but extremists pushed the movement to the fringe. At the same time, environmentalism became fragmented. Various factions debated the value of the natural environment and its relationship to human progress: Does nature exist to serve humankind, or vice versa? Does man have an ethical obligation to protect nature? If so, should he do so for his own benefit, or should he preserve nature for its own sake?

Today these questions have become scientific and economic queries about biodiversity, human health, and natural capital. Because we now know that we are depleting and polluting our most essential raw materials—such as water, forests, petroleum, and clean air—environmentalism has taken on a new personality in the twenty-first century. We've arrived at a point where philosophical and political issues can be put aside. We know scientifically that we must collectively come together to rethink the way things are done. To our credit, we've tackled other eco challenges: When scientists told us that the ozone layer—the part of the atmosphere that protects the Earth from the sun's harmful UV rays—was being depleted, humans stepped up to the plate and developed solutions to the problem. We can do the same for global warming.

Whether you've picked up this book for altruistic, ethical, or scientific reasons almost doesn't matter. You are part of a critical mass that is shaping the new wave of do-it-yourself environmentalism into a grass-roots social movement that has little to do with baggy hemp pants and tofu and everything to do with intelligent modern living.

> The last time that humanity was challenged to rethink the world, we came up with the Enlightenment, which served our kind very well up to now.
> —Susan S. Szenasy, editor-in-chief of *Metropolis*

The next industrial revolution—when the interests of technology, ecology, and commerce overlap—has already begun. Welcome to the bright green future.

circa 1750	Industrial revolution begins, changing the way things are made, distributed, and consumed. Natural resources are considered infinite.
1820	World population reaches 1 billion.
1872	America's first national park, Yellowstone, established.
1892	Sierra Club founded. Land conservation becomes a national topic.
1930	World population reaches 2 billion.
1956	"Peak oil" theory predicts we will run out of petroleum in the foreseeable future.
1960	World population reaches 3 billion.
1962	Rachel Carson's *Silent Spring* published, detailing how insecticides and pesticides affect human health. Modern environmental movement is born.
1968	First Whole Earth Catalog published.
1970	First Earth Day celebrated.
1970	U.S. Environmental Protection Agency established.
1973	First oil crisis takes place.
1974	First warnings of ozone layer damage are released.

1979	Second oil crisis takes place.
1980	First Whole Foods Market opens, in Austin, Texas.
1989	Exxon *Valdez* creates largest oil spill in U.S. history.
circa 1990	Recycling bins begin to appear on curbs across the United States.
1993	Patagonia begins making first fleece garments from recycled plastics.
1999	World population reaches 6 billion.
2002	*Cradle to Cradle* published, rethinking the problems and solutions of industry and dubbing the modern era the "next industrial revolution."
2002	U.S. Department of Agriculture creates "organic" food label.
2004	Kyoto Protocol ratified. International signatories pledge to cut carbon dioxide emissions by 5 percent overall from 1990 levels, between 2008 and 2012.
2005	Scientists proclaim this year the hottest on record.
2006	The documentary *An Inconvenient Truth* is released, making the threat of global warming evident to millions.
2008	You begin helping to save the world.

A TIME LINE OF ENVIRONMENTALISM

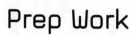

2

Prep Work
Getting Started

OUR ECOLOGICAL FOOTPRINT

You probably already know that everything you do has some kind of impact on the planet. But just how much impact do we have? How often we drive, how much meat we eat, the type of fuel we use to heat our homes, and so on, all contribute to the "footprint" we leave behind. The measure of how our lifestyles affect the Earth and its ability to regenerate resources is known as our Ecological Footprint. It can be calculated for individuals, organizations, cities, countries, or the entire world. Put simply, it is a calculation that works to ascertain planetary limits, like a spreadsheet of environmental checks and balances.

We all know that nonrenewable resources—such as oil, minerals, and ore—are finite and may someday run out. But if we deplete renewable resources—say fisheries, forests, and groundwater—faster than the planet can regenerate them, we will run out of these, too. Currently, our demand for the planet's renewable resources exceeds what it can supply by more than 20 percent, according to the Global Footprint Network. Put another way, the planet needs about fourteen months to regenerate all of the resources we use in one year. Luckily, the problem isn't insurmountable. By assessing where we're using too much and where we can cut back, we can return to a path of sustainability, where humanity's demands on nature are in balance with nature's capacity to meet those demands.

Our "carbon footprint"— the measure of how much carbon dioxide we emit— makes up about half of the

> To calculate your Ecological Footprint, go to www.ecofoot.org. To calculate your carbon footprint, go to www.safeclimate.net.

world's overall Ecological Footprint. Carbon dioxide—the main greenhouse gas responsible for global warming—is released any time we combust fossil fuels or make changes in the way we use land; an example would be clearing parts of the Amazonian rain

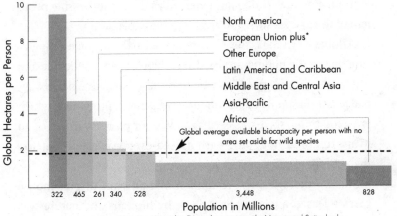

Ecological Footprint by Region (2002). The average person's Ecological Footprint is 2.2 global hectares (5.5 acres), while there are only 1.8 global hectares (4.4 acres) of biologically productive area per person available on the planet. This "deficit spending" is called overshoot. It is possible to exceed ecological limits for a while, but overshoot leads to the destruction of the ecological assets on which our economy depends, resulting in depleted groundwater, collapsing fisheries, CO_2 accumulation in the atmosphere, and deforestation.
Source: Global Footprint Network.

forest and converting them into agricultural plots. The other main anthropogenic greenhouse gases are methane and nitrous oxide, which are released primarily as a result of agriculture and things rotting in landfills.

Quantities of greenhouse gas emissions are often discussed in terms of carbon dioxide for the sake of making easy comparisons. When we say that eating an omnivorous diet creates emissions of 3,000-plus pounds of CO_2 per person per year, for example, that figure includes the equivalent amount of methane produced by such a diet.

GLOBAL WARMING: A PRIMER

While many ecological issues—loss of biodiversity, availability of clean water, endangered wildlife, for example—deserve urgent attention, global warming has emerged as the most pressing problem of the day. As such, it warrants some explanation.

Global warming is the increase in the Earth's average temperature due to the buildup in the atmosphere of carbon dioxide and other greenhouse gases. It's true that the atmosphere has always had a natural supply of greenhouse gases that capture heat, which is a good thing, since this is what makes our planet warm enough to inhabit, and not some barren, iced-over wasteland. But too many of these gases present a problem.

Before the industrial revolution, the amount of greenhouse gases released into the atmosphere by humans and nature was roughly in balance with what the Earth could reasonably store, or "sink." For example, a tree absorbs CO_2 during its lifetime, but releases it back into the air when it dies and decays. But when humans began burning tremendous amounts of fossil fuels, we created an imbalance of greenhouse gases, which become trapped in the atmosphere, acting like a blanket and heating up the surface of the planet.

Today, the amount of CO_2 in the atmosphere is more than 30 percent higher than preindustrial levels. This is almost certainly due to human actions, mainly agriculture, burning fossil fuels, and changes in land use such as deforestation. Today, there's more carbon dioxide in the atmosphere than at any time in the last 650,000 years. Geological history shows that changes in these levels—even small changes—are usually accompanied by significant shifts in global temperature.

> **20:**
> Percentage jump in the number of blizzards and heavy rainstorms in the United States since 1900.

Over the past century, the globe has heated up by about 1 degree Fahrenheit—a rate we haven't seen for more than a thousand years—with the most dramatic shift occurring over the past two decades. Day to day, one degree wouldn't even affect how you chose to dress in the morning. But that's the difference between weather and climate: Over the long haul, one degree is a really big deal. The last time we saw the polar regions heat up like this—about 125,000 years ago—sea levels rose between thirteen and nineteen feet.[1] That's enough to put New York, London, and Sydney underwater.[2]

While warming in some areas may appear advantageous in the short term—say, promoting longer growing seasons in various regions of the Northern Hemisphere—the overall negative effects will far outweigh these localized "benefits." Some countries in Africa, for example, may see a significant reduction in crop levels as soon

> **›$200 billion:**
> Global economic damages from weather-related disasters during the 1990s—four times the total losses reported during the 1980s.

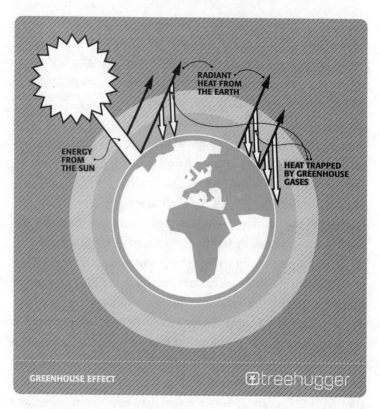

It's true that specific weather-related events such as the Northwest's drastically cold 2006/2007 winter season or Hurricane Katrina can't prove or dispel global warming. In the short run, these can be viewed only as anomalies. But emerging patterns—such as the increasing intensity of tropical storms—indicate that serious

as 2020. Sure, in the short term, most developed countries will be able to cope. But if current levels of greenhouse gas emissions are left unabated, by the end of this century we're likely to see more intense storms, pronounced droughts, rising sea levels, and widespread disease. More than one third of all species could face extinction.

change is already occurring. And in a study of nearly one thousand articles published on climate change in scientific journals, none questioned its validity.[3]

Some people believe that governmental policy addressing global warming will hurt our current carbon-based economy. This has supplied some big businesses and politicians with an excuse for not taking action. But creating well-designed cap-and-trade emissions programs, for example, could reduce compliance costs. Mandatory caps would spur technologies bound to jump-start the economy, not drag it down. Plus, the longer we wait, the higher the cost of cleanup will be. Wait too long and our economy—not to mention life as we know it—could collapse. It's not just about saving the planet—Earth itself will pull through—it's about keeping human beings off the endangered species list.

HOW TO USE THIS BOOK

Environmentalism isn't just about how humans affect nature; it's also about how the environment affects us. We are part of the ecosystems in which we live—not separate from them. Quality of life is a give-and-take. Our primary objective is to get you to think like a TreeHugger, and to provide a practical guide for living a healthier, more sustainable life.

Thinking like a TreeHugger is about making changes, not sacrifices. We believe that less is often more. That doesn't mean giving up the good life, but rather redefining it. We prefer quality to quantity and efficiency to waste. Modern environmentalism is about working smarter, not harder.

To that end, we have provided two lists of action items for each week. One, "Save the Planet in Thirty Minutes or Less," provides the simplest, most effective ways you can make a difference. Once you've mastered these basics, we hope you'll attempt some of the next steps on the other list, "So You Want to Do More," which shows how you can make your life even greener. An "Action" ticker shows how many pounds of CO_2 you can save annually by performing certain actions.

We encourage you to take on as many tasks as possible, but we also know that each suggestion won't fit everyone's personal lifestyle. So think of these activities as items in a toolbox: You've got a wide array of tools at your disposal. You decide which ones work best to get the job done. After all, the easiest way to start making greener choices is to evaluate your unique situation. Ultimately, that's what sustainability is all about: creating scenarios that continue to work over time.

3

Week One

Thinking Like a TreeHugger:
Reduce, Reuse, Recycle

Your mission:
Green your mind and the rest will follow.

Fifty years ago, most of what we consumed was produced near the places we lived. Today we buy even simple things such as drinking water from halfway around the world—and that adds up to a lot of trash as well as transportation emissions. The average U.S. citizen produces about 4.5 pounds of garbage every day, making us the industrialized world's leader in generating municipal solid waste. Germany and Sweden, by contrast, generate less than 2 pounds per capita each day. Think your household is small potatoes? Residential waste makes up 55 to 65 percent of the municipal solid waste stream.[1]

The good news is that this means that consumers and households can make a significant contribution when it comes to actualizing the TreeHugger mantra "reduce, reuse, recycle." As a nation, we've increased our rate of recycling fivefold in the past half century—a good start. But the sheer amount of stuff we toss has gone through the roof: In 1960, we generated 88 million tons of garbage. In 2005, that figure was up to more than 245 million—indicating that our reducing and reusing could use a little boost.[2] Our landfills are becoming heavy on the "fill" and neglectful of the "land." And running out of space isn't the only problem.

A THROWAWAY CULTURE

Things don't disappear when we throw them "away." Though the landfill may be out of sight, it shouldn't be out of mind. When everyday items such as plastic CD cases, grass clippings, and last night's leftovers end up in the dump, they either decompose, releasing methane—a greenhouse gas about twenty times more potent than carbon dioxide—or are incinerated, releasing CO_2. In fact, landfills are the number one source of anthropogenic methane emissions, comprising about 25 percent of the total, according to the U.S. EPA.

3:

Number of hours a television set could be run from the energy saved by recycling one aluminum can.

Degradation occurs slowly: Since sunlight can't penetrate most of what's stuck in the trash heap, things can't break down. Even readily biodegradable items such as cabbages, carrots, and newspapers have been found in landfills after more than thirty years.[3] The trucks hauling off our unwanted refuse produce a significant amount of CO_2 themselves. And getting rid of all this stuff indicates that we're replacing it with new products, which, of course, require materials and fuel to create. When we begin to understand that there is no such place as "away," we begin to realize why reducing, reusing, and recycling—in that order—are so important.

PAPER, PLASTICS, AND YARD WASTE

Even in the digital age, the most common thing found in garbage dumps is paper, accounting for an average of 35 percent of landfill contents overall. Yard waste comes in at a distant second, at 13 percent, followed by food scraps and plastics, both hovering around 11 percent. The remainder is a conglomerate of metals, leather, rubber, textiles, wood, glass, and a smattering of other materials.[4]

Of all the *types* of stuff we toss out, containers and packaging make up the biggest chunk, at about a third.[5] According to some accounts, more than 90 percent of materials extracted to make durable goods in the United States shockingly become waste almost immediately.[6] Think about it: Whether it is a candy bar, a CD, or a telephone you've just bought, you undoubtedly toss a wrapper or container into the trash even before taking a bite or making a call.

40:

Percentage of energy saved by recycling newsprint rather than creating it new.

Even more astounding, the physical product you've bought contains, on average, only 5 percent of the raw materials it took to make and deliver it.[7] And if you carried your purchase home in a plastic bag, the extent of our throwaway culture becomes even more evident: A single plastic bag is used for an average of twenty minutes, but takes about five hundred years to degrade in a landfill. (Some scientists believe that plastic bags take as many as one thousand years to degrade, but since they haven't been around that long, no one can know for sure.) Worldwide, hundreds of millions of bags are distributed each year; 100 billion of them in the United States. The United Nations Environment Programme estimates that globally, people go through 16,000 plastic bags per second.[8,9]

E-WASTE

In our newer, faster, need-it-now culture, the outpouring of gadgets and gizmos—not to mention the batteries that run them—is also beginning to pile up. The United Nations Environment Programme estimates that 20 million to 50 million metric tons of electronic waste, known as e-waste (that's iWaste to Apple users), is dumped around the world each year—65,000 tons alone in the United States.[10] An estimated 75 percent of our obsolete electronics—about 500 million gadgets—is stored unused, posing problems for future disposal.

Much of the flotsam doesn't stay put. Like many industrialized nations, the United States, which properly recycles only 10 percent of its e-waste,[11] exports much of it each year, especially to Asia. Though developing nations are hungry for large quantities of secondary raw materials, refurbishment facilities don't necessarily meet adequate environmental or health standards. Still, it can be cheaper to ship unwanted computers to, say, India, than to recycle them here at home. While Europe and Japan have passed legislation making e-waste recycling mandatory, only a handful of U.S. states have done the same.

In addition to losing literally tons of valuable materials such as copper and gold when we don't recycle gadgets properly, we build dozens of electronics components using at least some kind of toxin. Computer circuit boards and CRT (cathode-ray-tube) monitors and television sets contain lead and cadmium. Cell phones—more than 140 million of which are discarded in the United States every year—contain mercury and arsenic. The cables that hook all of our gadgets together? Many are sprayed with brominated flame retardants. And though mercury has been phased out of batteries, these small wonders are responsible for a large portion of our waste stream's toxic heavy metals. All of these toxins can leach into the soil and groundwater, poisoning wildlife and the drinking water supply, which can then lead to problems in humans such as cancer and neurological disorders.

The good news? Much of our electronic garbage can be recycled. Copper can be recovered from cell phone chargers. Wireless handsets yield plastics. Circuit boards contain gold, silver, palladium, and other metals. Batteries can be recycled for their nickel, iron, cadmium, lead, and cobalt. Many municipalities and some retailers will collect these and other hazardous waste materials for recycling or proper disposal. Consumer responsibility in getting these unwanted things to the right spot is paramount.

BATTERIES

Billions of batteries are purchased in the United States each year. And though they make up less than 1 percent of municipal solid waste, they are responsible for a large portion of toxic heavy metals. In 1996, the use of mercury in batteries was phased out, proper recycling methods began, and labeling that included disposal information became required. Although rechargeable batteries contain heavy metals such as nickel-cadmium, these are preferable to the single-use kind, since it means less stuff entering the waste stream.

Disposing of Hazardous Materials

Rechargeable batteries should be properly disposed of at the end of their useful life.

Any other household hazardous waste accumulated (paint, paint thinners, cleaners, oils, pesticides, etc.) should be disposed of according to municipal guidelines. Some items, such as motor oil, paint cans, and even paint, can be recycled—you'll have to take them to a local recycling center, though; don't just throw them in a curbside recycling bin. Visit earth911.org to find a venue in your region.

THE PROBLEM WITH PACKAGING

When it comes to bottled water, "the trade crisscrossing the globe defies common sense," reports the United Nations Environment Programme. France, for example, is the largest exporter of bottled water, but it's also one of the biggest importers. In addition to the energy it takes to move pallets of water around, bottles obviously use up a lot of plastic, which is typically derived from fossil fuels. This packaging is used for only a very short time, but it takes ages to decompose. Likewise, shampoo bottles, salad containers, and plastic baggies last for hundreds of years, while the useful products inside them last just months, weeks, or sometimes mere minutes. Most of today's packaging is seriously over-engineered.

185,000:
Number of steel-frame homes that could be built from material recovered through recycled packaging in 2005.

Luckily, this is beginning to change—albeit slowly. In 2003, for example, the Sustainable Packaging Coalition had just nine members; it now has nearly one hundred, including international bigwigs such as Starbucks and Estée Lauder, which have the purchasing power and visibility to create shifts within industries. In 2006, Wal-Mart announced it wants to reduce packaging by 5 percent by 2013, which could reduce CO_2 emissions by more than 735,000 tons and save the company $3.4 billion.[12] How? Less packaging means that more stuff can fit into fewer distribution trucks *and* that products take up less space on precious store shelves, leaving more room for other merchandise.

Action: Annual Savings in pounds of CO_2 • Reduce the garbage you generate by

THE ART OF PRECYCLING

It's not just big business that can help enforce change. The single best way to help reduce waste, unsurprisingly, is simply to consume less stuff. For the things we do need to buy, choosing the options with less or recyclable packaging or buying in bulk is helpful. Could you buy loose cucumbers instead of those wrapped in cellophane? Could you reuse screws in your toolbox instead of buying new ones? Do you really need another T-shirt? Could you download an album instead of buying the CD? Could you buy concentrated laundry detergent instead of liquid? (It uses less packaging per unit and also means more of it can be transported using less fuel.) Could you eke out one more year from that toaster?

The notion of reducing the amount of stuff that ends up in our trash bins before we even make a purchase is known as "precycling." In terms of climate benefits, the U.S. Environmental Protection Agency notes that when it comes to our garbage, waste prevention—rather than waste management—is the best way to reduce environmental stress. Reuse is the next best approach. Could you reuse a glass food jar as a container for leftovers? Could you furnish your guest room with a bed from a thrift store? Could you donate that lawn mower instead of taking it to the dump? These actions are simple and obvious, but each helps reduce waste and greenhouse gas emissions and saves natural resources, since fewer goods are manufactured, shipped, and landfilled.

> The average American home may contain as much as
>
> **100**
>
> pounds of household hazardous waste at any time.

THE BENEFITS OF RECYCLING

Twenty years ago, only one curbside recycling program existed in the United States. By 2005, that number had increased to nearly nine thousand, according to the EPA. Americans currently recycle about 32 percent of our total waste—not bad by world standards—though the EPA estimates that we could up that rate to 75 percent.

By decreasing the need to extract, process, and transport virgin materials, recycling helps reduce pollution. Life cycle analysts—who use a complicated methodology to consider the impact of a product or process over its entire life, from extraction and processing of the raw materials to production, distribution, use and reuse, maintenance, recycling, and disposal—have deemed that in most cases, less energy is needed to manufacture products from recovered rather than virgin materials. According to the EPA, "When all the pieces of recycling are put together, the overwhelming conclusion is that recycling boosts the economy, conserves natural resources, and reduces solid waste."[13]

Furthermore, since recycling can also spare trees from being harvested, it benefits the planet's natural ability to absorb carbon dioxide and helps deter the loss of biodiversity. Some economists estimate the lost pharmaceutical value from plant species extinction in the United States alone to be almost $12 billion.[14] Translation: Recycle those Sunday papers; if everyone did, we'd save five hundred thousand trees *each week*. Reading the paper online, of course, would save even more.

Reusing aluminum, for example, is 95 percent more energy efficient than creating new products from scratch. Put another way, you could make twenty recycled aluminum cans with the energy

and steel cans: 166 • Recycle glass: 26 • Recycle plastics: 47 (Sources:

necessary to produce just one can from virgin ore. Recycling plastics uses just 10 percent of the energy it takes to make a pound of plastic from virgin materials.

In some places, the fiscal cost of recycling outweighs the economic reward, begging the question of whether recycling is "worth it." But many municipalities have found that improved recycling increases jobs and saves money. Pay As You Throw programs encourage recycling by charging residents for only the amount of trash they produce, rather than a flat rate, rewarding recycling. Someday we may even see garbage and recycling bins with electronic sensors that weigh our trash and charge us accordingly. When we consider that the manufacture, distribution, and use of products—and their ensuing waste—result in tremendous greenhouse gas emissions and consumption of natural resources, the real question becomes, can we afford *not* to recycle?

Recycling by the Numbers

Resin codes—those numbers you see on the bottom of plastic containers inside the chasing arrows "recycle" symbol—indicate what type of material a product or package is made from, and how recyclable it is. They tell you what you can and can't toss in your curbside bin (which varies by community).

 PET/PETE (polyethylene terephthalate)
Common uses: Also known as polyester. Commonly used in beverage bottles and other food and nonfood containers.
Recycling factor: High. Almost always accepted curbside.
Second life: Carpet fiber, textiles, fleece jackets, new containers.

Recycling by the Numbers (continued)

 HDPE (high-density polyethylene)
Common uses: Milk, bleach, shampoo, detergent, and household cleaning containers, as well as grocery bags and cereal box liners.
Recycling factor: High. Almost always accepted curbside.
Second life: Nonfood containers, decking, fencing, flowerpots.

 PVC/VINYL (polyvinyl chloride)
Common uses: Packaging (blister packs, shrink wrap), construction (pipe, siding, window frames, fencing, flooring), medical tubing, cable insulation.
Recycling factor: Low. Not accepted in many communities.
Second life: Pipe, gutters, carpet backing, packaging.

 LDPE (low-density polyethylene)
Common uses: Flexible container lids; dry-cleaning, trash, and bread bags; shrink wrap; beverage container coatings.
Recycling factor: Fairly low. Accepted in some communities.
Second life: Shipping envelopes, trash bags, compost bins.

 PP (polypropylene)
Common uses: Molded automotive parts, food containers such as yogurt cups and takeout packaging, medicine bottles, bottle caps.
Recycling factor: Difficult to recycle. Accepted in few communities.
Second life: Rakes, storage bins, shipping pallets.

 PS (polystyrene)
Common uses: Protective packaging, bottles, food containers, cups, plates, bowls, cutlery, CD cases, videocassette cartridges, coat hangers.
Recycling factor: Low. Not typically accepted curbside.
Second life: Light-switch plates, desk trays, protective packaging.

 OTHER (all other types or a mix of plastics)
Common uses: Bottles, oven-baking bags, and various packaging.
Recycling factor: Very low.
Second life: Can sometimes be made into bottles or plastic lumber.

SOURCE: American Chemistry Council

UPCYCLING

While recycling is on the upswing—and should be regarded as a bridge to a greener future—it's not a perfect solution to our trash woes. The problem with recycling and reuse is that in many scenarios it's simply a way of extending the time until a product eventually hits the trash heap. That is, most recycling is actually "downcycling"—it reduces the quality of the material over time.

Most materials and products today are designed in a linear, cradle-to-grave fashion—make, use, throw away—rather than to be disassembled or recycled. A single product might contain more than a dozen different types of plastic, plus a variety of other materials—a nightmare of a disassembly job, and the reason why it's often cheaper to produce new stuff made from virgin materials than to capture and reuse existing ones.

> **2.5:**
> Number of cans of soda the average employee consumes each day at work.

Furthermore, recycling itself requires energy and resources. While the recycling rate of newspapers, for example, hovers around an impressive 89 percent,[15] reprocessing paper typically requires tons of water and extensive bleaching and other chemical processes, which release dioxins and other pollutants. Like so many products, paper wasn't *designed* to be recycled. Recycling as we know it is a reactive process.

But what if we could make materials that were endlessly recyclable into the same quality of product again and again (and again)? What if paper were conceived to be endlessly recyclable into the same quality product, without causing harm to the environment? What if paper didn't come from trees, but from endlessly reusable polymers? And then, what if we printed on it with ink that contained no toxins, and that could be washed harmlessly off the page? In their seminal book *Cradle to Cradle: Remaking the*

Way We Make Things, chemist Michael Braungart and architect William McDonough call this idea "upcycling." Their book, made from a type of polymer "paper" described above, can itself be end-lessly recycled into a new book of the same physical quality.

250,000:
gallons of drinking water that can be polluted from a single quart of motor oil that seeps into the ground.

This "next industrial revolution" isn't science fiction; it's already begun. Products smartly designed from their conception to be easily disassembled or recycled already exist. Herman Miller's Mirra desk chair is made from a minimal number of parts, can be easily disassembled, and is 96 percent recyclable. Designtex's sustainable fabrics are safely biodegradable or can be recovered and reused. To make a dent in the 50 million diapers landfilled in the United States every day, gDiapers designed a flushable, compostable version that takes days or weeks—not hundreds of years—to break down.

USING WASTE AS A RESOURCE

Braungart and McDonough suggest that human endeavors should emulate nature, a system where there is no waste but rather a constant stream of elements created and effectively absorbed back into the system, where "waste equals food." Think of it this way: Have you ever heard of an ant dump or a coyote landfill? Of course not.

Applied to human endeavors, this type of closed-loop, or "cradle-to-cradle," design would beget products and packaging that constantly resupplied the system. Plastic computer casings could be made from cornstarch instead of petroleum and would be tossed onto the compost pile. Car emissions could potentially *improve* air quality. This idea doesn't apply only to manufactured products. Composting—enabling organic materials to break down and create new nutrients for the soil—is probably the simplest example of a closed-loop system.

Eco-myth:
Municipal recycling uses more energy than it saves.

Fact:
Reusing aluminum is 95 percent more energy-efficient than creating products with raw materials. Recycling plastics uses just 10 percent of the energy it takes to make a pound of plastic from virgin resources.

Looking at the system as a whole necessarily creates a shift to a new way of thinking. Instead of asking only "How can I manufacture this product more cheaply?" we begin to wonder "How will this process's effluents affect the ecosystem?" Instead of designing to put money in our pockets, we start to consider the "cost" of things in much broader terms. How do these products affect the whole system, both now and for generations to come?

Composting in Five Easy Steps

Composting may not be sexy, but it can help significantly reduce the amount of organic (as in "carbon-based") waste you're sending to the landfill. Plus, it's a cinch. (Apartment dwellers, you're not necessarily off the hook. VermiComposting—that's composting with bins of worms—is easier and less gross than it sounds. Use the end result—what gardeners refer to as "black gold"—on houseplants, or donate to a plant-rearing pal.)

1. Place your composting bin in direct sunlight. The base must either touch the soil, which allows microorganisms to pass from the ground into the compost, or you can use compost starter.

2. Collect food scraps, cardboard egg cartons, plant materials, and the like, which you'll add to the outside bin. (*Note:* Avoid meat

Composting in Five Easy Steps (continued)

and dairy, which attract pests and degrade at a different rate.) If your pile appears dry, add water. Too wet? Add sawdust, shredded newspaper, or something similar.

3. Technically, you should have somewhere between twenty-five and forty parts carbon (the harder stuff, such as woodchips and cardboard) to one part nitrogen (the soft stuff, such as orange rinds and grass clippings). But, really, who has time to do the math? We just make sure there's a lot more carbon-based goop in our pile.

4. Depending on the climate where you live, it can take as little as a few weeks or as long as a few months for the muck to officially become compost. When it does, use it to enrich the soil in your garden.

5. Marvel at nature at work.

EVOLUTION, NOT REVOLUTION

Blown up and applied to entire systems, cradle-to-cradle designs could spawn buildings that, like trees, create more energy than they consume and purify their own waste water, factories whose effluents are clean drinking water, and machinery whose every component can be upcycled, according to Braungart and McDonough. Without the need to extract and process raw materials for everything they build, the duo predicts, companies would save billions—perhaps even trillions—of dollars each year in materials recovery; pollution and waste would be reduced drastically.[16] "Profit" would be measured not only by financial gain, but by the "triple bottom line"—ecology, equity, and economy. We'd ask ourselves, "Is this product/service/system good for the environment? For other people? For my pocketbook?"

In a world of Wal-Mart "rollback" pricing and quarterly finan-

cial reports, this idea may
seem puzzling—even nuts.
But several companies—
such as floor covering
manufacturer Interface,
household products manu-
facturer Seventh Genera-
tion, and outdoor clothing
and gear company Patago-
nia—are already following
this model, and with suc-
cessful results both ecologi-

> Consider this: All the ants on the planet, taken together, have a biomass greater than that of humans. Ants have been incredibly industrious for millions of years. Yet their productiveness nourishes plants, animals, and soil. Human industry has been in full swing for little over a century, yet it has brought about a decline in almost every ecosystem on the planet. Nature doesn't have a design problem. People do.
>
> —Michael Braungart and William McDonough, *Cradle to Cradle: Remaking the Way We Make Things*

cally and economically. These companies aren't looking to make
the highest profit at *any* cost (though they're all doing quite well);
they're considering *all* costs when calculating their profits.

PRODUCT SERVICE SYSTEMS

How can we bridge the gap between the present and this futuristic
world of harmony? One way may be to embrace the concept of the
product service system, or PSS. The idea is to "dematerialize" consumption but still meet practical needs. As consumers, the theory
goes, we seek not the product itself (a washing machine, say), but
rather the functionality (clean laundry) that it offers us.

A public library is a classic example of a PSS—you have access
to all the books you want, without owning any of them. Public
transport and Laundromats are others. With cable TV service,
you get the programming, but the provider typically owns the
box. Zipcar, which gives members in a handful of cities access to
a shared fleet of vehicles, and Netflix, an online DVD rental service, are great examples of new economy PSSs. You can even

20,000:
Number of steel cans recycled
every minute in the United States.

lease solar panels this way through the Citizenrē REnU program. With fewer objects shared by more people, less stuff needs to be manufactured.

Think of it this way: When it comes to preserving your food, for example, do you care about your refrigerator, or do you care about dinner being fresh? Okay, so maybe you *do* care about having the latest Sub-Zero stainless-steel fridge. But if Sub-Zero leased you the *service* of using the refrigerator instead of selling you the equipment outright, they'd have good cause to keep it in working order and recycle parts while you'd always have crisp lettuce.

Ongoing maintenance is usually cheaper for a manufacturer or company than replacing an entire product or system; by contrast,

100 million:
Number of cell phones Americans
tossed out in 2006. Recycling them
would have saved enough energy to
power 194,000 homes for a year.

we all know that it's often less expensive to purchase a new version of even the most expensive appliance than to get it repaired—not a very eco-friendly solution. A product service system is what TreeHugger calls a win-win-win: You get the end result you need, the provider of the service makes money, and less stress is put on the environment.

BIOMIMICRY

Nature represents the ultimate closed-loop system, and a relatively new science known as biomimicry is looking to nature to solve human problems. Take the example of spider's silk. Compared with the synthetic fiber Kevlar, which is used in bulletproof vests, spider's silk absorbs five times the impact force and stretches 40 percent longer than its original length without breaking. On a human scale, it's so resilient it could catch a passenger plane in flight, according to the book *Biomimicry: Innovation Inspired by Nature,* by Janine M. Benyus.[17]

Unlike Kevlar, which is made using petroleum and sulfuric acid, spider's silk is naturally occurring, nontoxic, and readily biodegradable. Mimicking nature's version of throwaway items—making materials that are as strong as we need them to be but that return to a nat-

> Nature recycles everything.
> —Janine M. Benyus, *Biomimicry*

ural state once we are done with them—could ultimately help humans figure out how to run communities on sunlight and other endlessly renewable resources and, ultimately, make waste equal food.

Q & A

Architect and Author William McDonough on Closing the Loop

Can recycling as we know it be the cure for society's waste issues? Recycling is a critical act for the future. That should not be misunderstood. But in the typical model, things are downcycled. We need to recycle at the same level of quality, or upcycle things to a higher level of quality, where materials are getting better and better. In the cradle-to-cradle philosophy, things are returned either to a biological cycle—re-

Q & A (continued)

turned to the soil—or to a technical cycle. Downcycling is when materials get lost in the landfill.

What will the world be like when cradle-to-cradle design is the norm? We'll have an abundance of good things processed in ways that are beneficial to natural systems. We'll see a celebration of closed-loop cycles.

What's the biggest obstacle we currently face in reaching this goal? Design is a signal of human intention. It's how we do things. So I would say that bad design is the biggest obstacle. Over the next hundred years, we probably will figure out how to change things. My design firm, MBDC, offers a Cradle to Cradle Certification for products that helps designers take a look at things on a molecular level. We've found better replacement chemicals, and those will be integrated into products over time. It's very exciting.

If you could overhaul one service or system right now, what would it be? Energy. We need to create efficiency and large-scale energy savings wherever we can, but not at the cost of toxicity. We need renewable energy deployment at a high speed. Instead of carbon offsets, we need to drop the price of solar collectors by a factor of five or ten, for example.

In what sectors of industry do you see the most promise and leadership for achieving an eco-effective future? As an architect, it's thrilling to see how many people have adopted green building. There's a real desire to participate. I think we'll look back in fifty years and be surprised that it

Q & A (continued)

didn't come from industrial design or the energy sector, but from the building and architecture communities.

What are the most important things individuals can do to help set this revolution in motion? It starts with engaging discussion within the family, which then spreads it within the culture. Communities have power. When it comes to individual acts, it's things like composting, energy efficiency, and switching to and supporting renewably powered energy. The smart redesign of some packaging—particularly with some electronics and plastics—is notable. Encourage that with your purchasing decisions, and communicate with companies when that affected your decision.

You're fully immersed in the green revolution. Do you have any eco–guilty pleasures? I like to take vacations with my family. So the pleasure is in visiting places, but the guilt is in the carbon burden associated with travel.

William McDonough is founding partner of the architecture firm William McDonough + Partners and cofounder and principal of the design firm McDonough Braungart Design Chemistry.

Save the Planet in Thirty Minutes or Less

 reduces CO₂

♡ improves health

$ saves money

⊘ saves time

- Start using a reusable coffee cup and water bottle this week. 🔩

- Learn as much as you can about your local recycling and hazardous-waste programs. Earth 911 (www.earth911.org/recycling) can help. 🔩

- Cancel at least one print magazine or newspaper subscription this week or transfer it to a digital mode. Start reading more online. 🔩 $

- Consider what product service systems are available to you and take advantage of them. Get a library card, join Netflix, or use a Laundromat. 🔩 $

- Each time you go shopping this week, think about what you're buying, observe how things are packaged, and recognize how much waste each product will create. 🔩

So You Want to Do More

🔩 reduces CO₂

♡ improves health

$ saves money

⊘ saves time

- Keep America Green estimates that the average household has between three and five unused cell phones lying around. Scour your home for unused mobile phones, computers, and other electronics. Round them up and take them to a recycling center. 🔩

- Go paperless. Arrange that your credit cards, car loans, and other accounts provide statements online. Ask vendors and organizations to send information via e-mail rather than snail mail. 🔩 $ ⊘

- Invest in rechargeable batteries. 🔩 $

- Begin to think seriously about how you can reduce waste. Start a compost pile. Recycle more. Buy less. 🔩 $

- Round up all of the hazardous waste in your home and dispose of it properly. 🌀

- Use a mash-up of old paint instead of buying a new can for your next fixer-upper project. $

- Start reusing everything you can—a pickle jar for buttons, a chipped coffee cup for nails, a rag instead of paper towels. 🌀 $

- Buy your next bike, couch, television, book, and so on, at a tag sale or secondhand shop. 🌀 $

- Get up-to-the-minute tips and information on reducing, reusing and recycling at PlanetGreen.com. For news and global trends on this and a zillion other topics, visit TreeHugger.com.

4

Week Two

Eating Your Way Green:
Food and Drink

Your mission:
Be smarter than the grocery store.

According to researchers at the University of Chicago, the average American's diet creates one and a quarter tons of carbon dioxide emissions per person per year.[1] That means our food choices can have as much impact on the ecology as the cars we drive. So eating wisely is a TreeHugger's duty. Luckily, that doesn't mean you have to nosh tofu and bean sprouts three times a day. Still, while some choices are irrefutably smarter than others, when it comes to greening your diet, navigating the menu of choices can be a complicated proposition.

BENEFITS OF NONCONVENTIONAL FARMING

You probably already heard the buzz about organic foods and better health. Since no pesticides are used to produce them, those same toxins aren't present in the food that enters your body or in the waste stream of the farm that produces it. If that doesn't tempt your palate, maybe the fact that organics generally taste better will. Over the past decade or so, organics have graduated from natural food stores and made their way into the fluorescent-lit aisles of major grocery chains, and are quickly making their way into well-known name-brand foods. Maybe you're wondering how this happened. Or maybe you're wondering what took so long.

The term "organic" refers to a system of farming that maintains ecological harmony and promotes biodiversity by replenishing soil fertility without using toxic pesticides or synthetic fertilizers. Organic farming prevents topsoil from eroding, keeps toxic substances out of nearby water sources, and conserves energy. And since organic foods must be produced using no antibiotics, synthetic hormones, genetic engineering, cloning, sewage sludge, or irradiation (the use of X-rays to sterilize foods), they're better for our bodies. Using no artificial ingredients or preservatives gives organic foods a health advantage, and some scientists believe they may even be higher in nutrients than their conventional counterparts. All of which gives

> Food is something we have it in our genes to care about, and we have been severed from that caring for too long.
>
> —Janine M. Benyus, *Biomimicry*

organic farming pretty high integrity, which industrial agriculture, while a bastion of American ingenuity in its own right, lost somewhere between the mass industrialization of World War II and the invention of Tang.

By changing the kind of crop produced in each field each growing season—a method known as crop rotation—organic farming can actually build up nutrients in the soil without synthetic fertilizers, which require large amounts of fossil fuels to create and can cause an overabundance of nitrogen and phosphorous to be deposited in the ground, air, water, and our bodies. Organic farms also battle pests naturally, whereas the pesticides, herbicides, and fungicides used in traditional farming can accumulate in the body. Children are particularly susceptible to pesticides, which have been linked to cancer and developmental disabilities.

Since 2001, national standards require organic growers and handlers to be certified by third-party state or private agencies or other organizations accredited by the U.S. Department of Agriculture. Farmers who sell less than five thousand dollars a year in organic products don't have to be certified, but technically they are still required by law to meet the same standards, so don't disparage your local organic farmer if his butternut squash doesn't have a "USDA Organic" sticker on it. Products made from at least 95 percent organic ingredients can carry the government's seal. The remainder must be ingredients approved for use in organic products. Certified products labeled "100% Organic" are just that—made entirely from organic ingredients.

The "Dirty Dozen" Top Fruits and Veggies to Buy Organic

Washing and rinsing fresh produce is always a good idea, but while it may reduce levels of some pesticides, it does not eliminate them, warns nonprofit research organization Environmental Working Group. Choosing organic produce can reduce exposure to harmful chemicals.

EWG's "Dirty Dozen" lists the foods with the highest levels of pesticides, starting with the worst.

1. Peaches
2. Apples
3. Sweet bell peppers
4. Celery
5. Nectarines
6. Strawberries
7. Cherries
8. Lettuce
9. Imported grapes
10. Pears
11. Spinach
12. Potatoes

Organic Farming and Global Warming

In addition to all of its health benefits, organic farming is also a lot easier on the Earth when it comes to global warming. Since organic farmers use no-till methods—that is, leaving soil and crop residue in the ground rather than digging it up—they release less

CO_2 into the air by sequestering it in the earth. According to the Organic Trade Association, organic farming uses 50 percent less energy overall than traditional farming; smaller-scale organic farms use 60 percent less fossil fuel per unit of food than conventional industrial farms. In short, organic farming is meant to ensure that the people doing it are stewards of the land.

From Farm to Tap

As chemicals from fertilizers and pesticides are absorbed into the ground, they end up in the groundwater and then in lakes, rivers, streams, and eventually the ocean. These toxins can even get into your drinking water. Nitrates, arsenic, lead, mercury, cadmium, chromium, and dioxin—which have been linked to cancer and a host of other health problems—have all found their way from the farm to the tap. And when unnaturally high amounts of nitrogen make their way into coastal waters, it contributes to the production of hypoxic, or low-oxygen, areas, known as "dead zones." Nearly 150 of these—as small as one square kilometer and as large as seventy thousand—exist in the world's oceans, where marine life is literally suffocated.[2]

THINK GLOBAL, SHOP LOCAL

We've established that growing organic food causes fewer greenhouse gases than traditional farming. But what happens when your organic asparagus is shipped from Chile to California, for example? The food we buy travels an average of 1,500 miles to get to our plates—not an insignificant distance. As your dinner racks up "food miles," it's also accumulating a larger ecological footprint due to the energy required to store and transport it across the country or around the globe. As a result of its time in transit, your asparagus probably also requires a lot more packaging to keep it

in good shape while on the move. Enter the virtues of locally produced, seasonal food—organic or not.

That's where farmers markets and community-supported
agriculture, or CSA, come in. The popularity of farmers markets
has skyrocketed over the past few decades; in 2004, there were

about 3,700 in the United
States, a more than tenfold increase since 1970.[3] When you
buy into a crop share from your
local CSA—usually available
even in cities—you invest in the
farm itself, receiving a portion
of the seasonal bounty at regular intervals, typically weekly.
While areas with shorter growing seasons will have a greater
variety during warmer months,

> In Spring 2007, British grocer
> Sainsbury's introduced designer
> Anya Hindmarch's reusable
> cotton tote printed with the
> phrase "I'm Not a Plastic Bag,"
> designed to encourage shoppers
> to decrease their reliance on
> disposable plastic bags. All
>
> # 20,000
>
> of them sold out within an hour.

even in New England some CSAs manage to provide meat, eggs,
and cheese pretty much year round.

Whether or not it's organic, food from CSAs and farmers
markets provide buyers with local, minimally processed and
packaged foods. Plus, farms selling directly to consumers are
usually smaller, so even if they're not organic, they typically use
less energy and fewer pesticides and herbicides than industrial
farms do.

Direct selling is also a better deal for the farmer: Instead of getting the customary 8 or 10 percent of the profit conventional
farmers receive, he gets all of the profit. That also means that more
of your money benefits and circulates in your community: The

Action: Annual Savings in pounds of CO$_2$ • Cut your beef consumption by one

farmer buys a coffee, the coffee shop owner gets his car fixed, the mechanic goes to the farmers market, and so on.

Of course, it's highly likely that you won't find everything you need or want through these venues, meaning you'll still need to make a weekly trip to your big-box grocery store. Some environmentalists argue that the gas you use making trips to different places to acquire various ingredients outweighs the food miles of simply shopping at the supermarket. And growing your own organic food, of course, would make you the ultimate TreeHugger, but not everyone's got a green thumb, the backyard in which to use it, or the desire or time to do so. It would be incredibly difficult—if not virtually impossible—for the average consumer to calculate the carbon costs of each food decision he makes. The decisions you choose should be based on what makes the most sense in your individual situation.

Paper versus Plastic

Life's all about choices—sushi or chimichangas? kelly green or chartreuse?—but one question continues to stump millions at the checkout counter each day: paper or plastic? Let's start with plastic, that petroleum-based scourge of the seven seas and breezy alleyways everywhere. Americans zip through more than 100 billion bags every year. Their destiny as bearers of many things may be short-lived, but plastic bags can look forward to a long retirement—possibly up to a millennium—before they decompose. That's if the choking hazards don't get mistaken for food by hapless marine life first.

Paper bags may be made from renewable trees, but they're hardly benign. According to the EPA, these sacks generate 70 per-

---

Paper versus Plastic (continued)

cent more air pollutants and fifty times more water pollutants than plastic bags. Not only does it take four times the energy to make a paper bag, it also takes 91 percent less energy to recycle a pound of plastic than it takes to recycle the equivalent in paper. The TreeHugger's solution? Boycott the lot, and flash a reusable shopping tote.

ORGANIC GOES MAINSTREAM

Traversing the aisles of your grocery store, you'll probably notice that organics are popping up in some unexpected places. Everything from Kraft string cheese to Ben and Jerry's ice cream is suddenly showing up with an organic label. As consumer awareness about the benefit of organics increases and retail outlets such as Whole Foods work to meet demand, sales of organic foods have soared, hovering around $16 billion in 2006.[4] This must be good news, right?

Some worry that "organic" may begin to lose its value as big-box stores such as Wal-Mart and Whole Foods begin to take organic mainstream. First, there's the problem of size: As industrial farms go organic and some organic farms grow into big industrial farms, smaller growers tend to get pushed out of the picture. This leads to a problem of scale: When big farms buy up the produce from little farms, the goods are processed and packaged more and transported farther. The connection between farmer and community withers, and the organic yogurt you thought was coming from Vermont is actually being made from powdered milk that came from Europe and then turned into yogurt back in Vermont. Not necessarily a step in the right direction.

More worrisome, there is also some concern that the big money behind "big organic" could be used to lobby the government to relax current USDA standards. On the other hand, organic is organic, and that can't be all bad. The Pew Center on Global Climate Change reports that greener changes in agricultural practices could reduce U.S. greenhouse gas emissions by one fifth, and that means that having big players in the organic market could be not only good business but good for the planet.

BREAKING THE BOTTLED WATER HABIT

Though two thirds of our planet is covered in water, less than 3 percent of it is fresh. Of that, only 0.003 percent is available to us; the rest is frozen into glaciers or ice caps, or is too deep in the earth to retrieve, according to the Rocky Mountain Institute. (And no, the fact that global warming might be melting those spots still isn't a good thing.) Unfortunately, of that tiny amount, more and more is becoming polluted.

More and more water is also ending up in plastic bottles, which is also disconcerting. For starters, many reports have shown that some bottled water may not be any better for you than what comes out of the tap. Worse, all of those containers have to come from somewhere; typically that means fossil fuels. The bottles are then transported around the globe. Apparently we're a thirsty lot: Americans buy nearly 30 billion single-use plastic water bottles annually and, despite being recyclable, the majority—or, more precisely, about 845 per second—end up in the trash heap.[5]

Cutting back on the bottled water habit (and other drinks) is essential to reducing your footprint. The best solution is a reusable water bottle that you can fill up at home and on the go. Most tap water is perfectly safe. If you feel that it's necessary to treat the water where you live, a tap-mounted filter or filtered pitcher, such as those made by Pur and Brita, extract toxins such as chlorine, mercury, and lead from water. Whole-house filters use technologies such as carbon filters, UV filters, and reverse osmosis to purify H_2O, but they can be pricey and will add to your home's energy load.

For bottles, stainless-steel and aluminum bottles are good choices; but anything reusable will do. One-time-use water bottles, which are made from resin type-1 plastic, are safe to reuse; if they become cracked or scratched, however, it's probably time to send them to the recycling bucket.

Because plastic bottles made from polycarbonate have been found to leach the chemical compound bisphenol-A, or BPA, some people prefer to avoid them. Polycarbonate, which is used in some drinking and baby bottles, is a type-7 plastic (though not all type 7 is polycarbonate). Though exposure from these sources is likely very small, some people and even retailers are avoiding them altogether, such as Mountain Equipment Co-op, which has removed all products containing BPA from their stores. If you'd prefer to err on the safe side, you can look for plastic bottles made from polyethylene or polypropylene (resin types 1, 2, or 5). Tempered glass is another option.

In 2004, Americans drank nearly

7 billion

gallons of bottled water.

Though the human dangers of BPA are not clear, it has been found to disrupt hormones in mice. Some scientists and the plastics industry claim that levels of the chemical that leach from plastics are safe; others are convinced that BPA can lead to health issues such as uterine fibroids, infertility, and behavioral problems in children. A more likely source of BPA is the linings of some food containers, such as certain aluminum cans. High levels of BPA have been detected in canned liquid baby formula; to be safe, use powdered versions.[6]

A NOTE ON COFFEE

Coffee is a shade-loving, tropical rain forest plant grown in fragile soils that are generally quite susceptible to erosion. Since rainforests have enormously high levels of biodiversity, with the farming of great coffee comes great responsibility. Coffee plants naturally grow under the canopy of taller trees. Hence the term "shade-grown" coffee—an indication of a product that's better for the planet—which refers to a crop that is grown in a way that more closely resembles the natural state of the jungle and so has less negative impact on its surroundings. Organic shade-grown coffee is even gentler on the Earth.

Irresponsible coffee farming can pose a serious threat to rainforest ecosystems and lead to deforestation, one of the greatest contributors to global climate change. Massive losses of rainforest acreage have a twofold damaging effect on the atmosphere: As trees are cut down, they both create more CO_2 and capture less. Deforestation also eliminates habitats for native wildlife and migrating birds.

Gidon Eshel and Pamela A. Martin, University of Chicago; EU Carbon Calculator.)

Is Starbucks Eco-friendly?

It's virtually impossible to talk about coffee without mentioning Starbucks, and you're probably wondering if your four-dollar latte is helping or harming the planet. As the second most widely traded commodity in the world, coffee has global social and ecological impacts. (It also costs far more per gallon retail than gasoline.) Starbucks touts a mission of environmentalism and social justice—purchasing shade-grown and organic coffees, paying fair-trade prices for beans, and paying employees relatively decent wages. It has also invested in renewable energy and recycled-content cups.

Critics, however, claim that the company's actions smack of greenwashing—projecting environmental benefits for PR purposes that belie its true actions—that Starbucks's fair-trade practices are not as humanitarian as the coffee giant would have customers think, and that its organic and fair-trade product offerings are too slim. While those accusations may or may not be true, few could argue that Starbucks is doing a worse job than its major competitors.

THE OMNIVORE'S DILEMMA

While enjoying the occasional filet mignon is understandable, eating less meat—especially less red meat—is the smarter environmental choice. Raising animals for food increases carbon emissions because of the fuel it takes to cultivate, harvest, and ship animal feed; transport animals to slaughter; and then process, package, store, and transport that protein-based portion of dinner.

In fact, the average meat eater is responsible for the equivalent of about 3,000 pounds more carbon dioxide emissions per year than a vegetarian, say researchers at the University of Chicago.

That's the difference between, say, driving a traditional Toyota Camry and a hybrid Prius. A vegan diet—one that includes absolutely no animal products at all, not even milk, eggs, or honey—is even easier on the planet.[7] Whether to go veg is up to each individual, but eating less meat is an easy step toward decreasing your footprint.

As anyone who has ever driven in a car with the windows down past a field of cows can attest, ruminants (mammals such as sheep and cows that chew their cud and have multichambered stomachs)—particularly beef cattle and cows—are pretty stinky. The gassy aroma comes from the animals' waste, which produces methane. It also takes land—and lots of it—to raise cattle and other grazing animals. In some regions, particularly areas of South America, this has been a major factor in deforestation. In others, grazing animals have devastated vegetation. In many cases, the same amount of land could produce vegetables and grains to feed many more people than beef or other meat would.

Happy Cows, Happy Milk

If your beef with meat and other animal products is an ethical one based on compassion for animals, you may be pleased to know that organic husbandry tends to institute practices that provide animals with living conditions that may be more in line with your ethos. Unlike some of their industrially raised counterparts, organic livestock spend time outdoors, have access to grass and pasture, exercise more, and eat only organic feed. Organically raised ruminants, which are natural herbivores, are never fed animal byproducts.

Since antibiotics cannot be given to animals whose meat is to be sold as organic, farmers have a huge incentive to create living conditions wherein animals are less likely to get sick in the first place. In traditional farming, antibiotics are given to livestock that are not sick, creating bacteria that become resistant to the medicine. Meats labeled "natural," by the way, are minimally processed, have no additives, and are often—though not necessarily—antibiotic- and hormone-free. "Natural" is not the same as organic or grass-fed, however.

Grass versus Grain

Although most beef raised in the United States is fed a hearty diet of corn, the staple is not a part of the animals' natural diet. Corn found its way into the trough decades ago, mainly due to the fact that U.S. farmers produce an overabundance of it, making it very cheap and widely available. Today, foodies nationwide maintain that grass-fed beef is more natural and better tasting—and that its production is better for the animals.

> **Eco-myth:**
> Eating organic and grass-fed beef has a significantly lower environmental impact.
>
> **Fact:**
> Regardless of how it is raised, beef is an energy- and water-intensive food to produce. Diets lower in any kind of meat create a smaller footprint.

Because grass-fed beef can use more water and land in order to feed the animals, certain types of grass can alleviate this problem; plus, grass-fed beef is often leaner—and therefore healthier—than its grain-fed counterpart.[8] Carefully managed grazing on permanent pastures may also have ecological benefits, such as reductions in greenhouse gas emissions, pollution, and overall energy consumption. (Poorly managed grazing can have the opposite effects.) When shopping, you can look for the grass-fed label; the American Grassfed Association also maintains a list of milk and beef producers that follow their standards.

Tackling Issues with Fish

Fish can be an essential element in a healthy diet, providing low-fat protein and omega-3 fatty acids. But not all fish are created equal. For example, doctors recommend that pregnant women not eat any tuna, swordfish, tilefish, or shark at all, due to their extremely high levels of mercury and other environmental toxins, which are detrimental to the health of developing babies, and not so great for the rest of us, either.

Mercury and other pollutants, such as PCBs (polychlorinated biphenyls) and dioxins, come from commercial endeavors such as coal-fired power plants; the production of paper, plastic, and pharmaceuticals; and water treatment processes. Exposure to mercury can lead to memory loss, tremors, and other health problems, while exposure to dioxins and PCBs may lead to certain types of cancer. So how can you avoid getting sick without going hungry?

Eating lower on the food chain—that is, choosing foods that consume fewer other foods, helps avoid that risk. A carrot, for example, is very low on the food chain, while a shark is very high. Humans are pretty much at the top. When it comes to fish, one that has been around for a while and eats lots of other fish, such as tuna do, will have collected many toxins in its body, in a process known as bioaccumulation. A tiny herring, by contrast, which eats only plankton, will trap fewer toxins, therefore passing fewer on to you.

Fish farming, or aquaculture, is a mixed lot. In some cases, fish farming can save wild species from being overharvested. On the other hand, some farmed fish live like cattle in feedlots; that is to say, in very tight quarters. As a result, they need antibiotics to ensure the health of the group. As in land-based livestock, this promotes the resistance of antibiotics in the fish, and eventually in

Cheers for Beers

Impressed with the recent rise of the farmers market? Get a load of organic beer sales: a 40 percent increase since 2005 has the frosty beverage tied with organic coffee as the fastest-growing organic drink. We

THE BEER	MADE IN	WHY WE LOVE IT
St. Peter's Organic English Ale	Suffolk, England	Although we generally advocate eating local, we can't resist this crisp, refreshing brew. Bonus: Cool oval bottle.
Orlio Seasonal Organic India Pale Ale	South Burlington, Vermont	Strong but delicious, with a distinct taste of caramel and spice.
Samuel Smith Organic Ale	Tadcaster, England	From one of the oldest remaining independent breweries in England; this delicate ale features subtle fruit flavors.
Long Trail Blackbeary Wheat Seasonal Ale	Bridgewater Corners, Vermont	At Long Trail, environmental considerations inform every decision, making the hint of blackberry in this beer all the sweeter.
Cru D'Or Organic Belgian-style Abbey Dubbel	Fort Bragg, California	Originally brewed for Whole Foods Market's twenty-fifth anniversary; this deep brown brew is reminiscent of Chimay.
New Belgium Fat Tire Amber Ale	Fort Collins, Colorado	Can wind power and green architecture make beer taste better? Turns out the answer is yes.

think hops and barley grown sans pesticides taste terrific, and we also tip our glass to breweries using alternative energy, recycling water, and conserving natural resources. Find ones as close to home as possible, pop the top, and eco-imbibe!

PAIRS WELL WITH	PERFECT FOR
Seared sea scallops (farmed)	A sunny Saturday barbecue
Pumpkin pie	A Halloween party
Alaskan wild grilled salmon	Enjoying a sunset
Organic chicken fingers	Après ski
Local mild cheeses	That beer goblet you've never used
Tex-Mex	Hanging out at the pub with friends

humans, too. Waste from fish farms can also end up contaminating waterways.

The USDA doesn't have an officially regulated organic label for fish, though you may see the term on some packaging, but fish raised without antibiotics are usually labeled as such. Wild fish can be a good alternative, as long as the species or location isn't being depleted. (Monterey Bay Aquarium's Seafood Watch site offers handy pocket guides to which species are being overfished.)

GENETICALLY MODIFIED FOODS

Could genetically modified foods be the answer to combating world hunger? Or do they have the potential to wreak Dr. Moreau–style havoc on the land and sea? The answers are unclear, which is why we'd prefer that genetically modified organisms, or GMOs, be kept out of our foods, or at the very least that foods containing GM crops be labeled as such. Though organic foods never include any genetically modified organisms, traditional foods can.

The term "genetic engineering" refers to technologies that change the genetic makeup of cells and move genes across species. Unlike hybrid varieties of, say, tomatoes, which are bred using strains of different plants capable of reproducing in nature, genetically engineered organisms manipulate genetic material and biological chemicals from different species that could not mix naturally. Via genetic engineering, you could, for instance, cross a cow with a chicken, or a hog with an iris. This is a radical departure from traditional plant and animal breeding. To date, millions of acres of corn, soybeans, and cotton in the United States have been grown from seeds that have been altered to resist insects, weeds, and even herbicides. Because corn and soy are so widely used in food processing, small amounts of engineered ingredients show up in a majority of processed food products.

Those in favor argue that drought- and pest-resistant GM crops could help solve world hunger. But the United Nations Food and Agriculture Organization estimates that the world can produce enough food to meet global demand in the year 2030 without the use of GM crops. Current problems have more to do with poverty and allocation of resources than actual food production.[9] Furthermore, some experts believe that these crops may not be as pesticide-resistant as some scientists and farmers would like to think. Others worry that pollen and seeds from GM crops could be carried to neighboring fields by the wind and could then infiltrate surrounding areas, creating hyper-resilient "superweeds"

Know the Code

According to the Pew Initiative on Food and Biotechnology, 75 percent of Americans want to know if our food contains genetically engineered ingredients. Until we get our wish, here's a little trick you can use to decipher the origins of your meals. If the PLU—that number under the bar code—starts with a nine, you're holding an organic food product. Conventional fruits and veggies begin with a four. If an eight begins the sequence, you've got genetically engineered food headed to your plate.[10]

that could dramatically change ecosystems and significantly reduce biodiversity. The problem is that we just don't know what long-term effects GM crops will have on human health or the environment. Could there be a place for them in the future? Maybe. But until we know that they're not devastating, we'd prefer to take a rain check.

No genetically modified animals are currently allowed into our food supply, though engineered fish are under consideration. Cloned animals, however, are likely to be introduced into our food stream; their offspring are already allowed. Successful clones may be scientifically "normal," but experts at the watchdog group the Union of Concerned Scientists believe that "a small possibility remains that some [clones] may harbor subtle genetic defects that could render them unsafe for consumption... since most attempted clones are grossly defective and are lost early in development."[10] The UCS believes that the FDA should let people know if their meat and milk come from cloned animals or their offspring—and we do, too. Certified organic food, however, cannot contain any ingredients derived from cloned animals.

Q & A

Alisa Smith and J. B. MacKinnon, Creators of the 100 Mile Diet, on Eating Locally

What is the 100 Mile Diet?

AS: In the simplest terms, it means that all the food you eat is grown or produced within one hundred miles of your home. Most of what we ate during our yearlong experiment of eating local ended up coming from within fifty miles.

How did you come up with idea? Why do it?

AS: We had heard the statistic that most food travels a minimum of 1,500 miles from farm to plate—that seemed outrageous. Then one weekend we were at our wilderness cabin, and we ran out of food. There's nowhere to shop, so we caught fish and collected mushrooms. It was the first time we knew everything about where our food came from.

JBM: The freshness of that meal was what struck me. We took a lot more pleasure in our food.

Have your taste buds been spoiled?

JBM: Restaurants seem a lot blander to us now. We realize that we could eat a much better meal at home for minimal effort.

Q & A (continued)

What were the hardest things about this lifestyle change?

JBM: It takes a lot of time to eat this way, at least at first. We had to spend more time to find growers and producers and to grow our own food.

How long did it take you to get into a groove?

AS: About three months. That was how long it took us to find locally grown wheat. The flour thing was really disruptive.

JBM: For us, this was a total revolution. Before, 90 percent or more of our food came from nonlocal sources, and was often premade. Local eating is more provisional. We had to completely relearn how to shop and eat.

Did your health suffer?

AS: We felt very healthy. We had canned and frozen a range of foods when they were in season, and were eating those throughout the winter. But we did wonder if we were getting malnourished, so we called a nutritionist. She laughed at us!

JBM: If you compare a strawberry picked at its peak then immediately frozen to one that you buy "fresh" at the grocery store in the middle of winter, the preserved one is much healthier. The other sheds nutrients as it's shipped across the country.

What were the biggest rewards?

JBM: This went beyond reducing our CO_2 footprint. The most important aspect became reconnecting with the landscape. One of my biggest satisfactions was realizing the power we have to shape the system. Because people are now

Q & A (continued)

asking for local food, they're changing how grocers stock their stores. We have several farmers growing wheat here now because they see a demand.

AS: It's a powerful message. I'd also never thought much before about how things taste when they're ripe. I was amazed at how much better the food tasted.

What are the simplest ways to begin incorporating local food into our diets?

JBM: Switching to local honey, farm-fresh eggs, and meat works in almost any geography. Track down the nearest farmers market—a locus for connecting with your community and your food chain.

AS: Start with a 100 Mile potluck, where each person focuses on just one dish.

How much regional food do you eat now?

JBM: We eat about 80 percent local, supplementing with certain other foods from around the world.

Alisa Smith and **J. B. MacKinnon** are the authors of *Plenty* and maintain the website 100 Mile Diet.

Save the Planet in Thirty Minutes or Less

reduces CO_2
improves health
$ saves money
saves time

- Buy at least three organic vegetables or fruits this week. Since organics tend to be a bit pricier, don't worry about restocking your entire pantry all at once. You'll also be "voting" with your dollars by creating demand.

- Research farmers markets and CSA farms in your community by logging onto www.localharvest.org. Ask: Do they operate year-round? Is it feasible for you to shop with them?

- If you're an omnivore, eat one to two fewer servings of meat this week.

- Whether you pick something new or dedicate something you've already got, select a reusable bag or two for toting your groceries. If you typically drive to do your shopping, leave a bag in your car. Or leave it by your doorway—just commit to using it. Even if you buy more groceries at one time than your bag can hold, you're still reducing the number of paper or plastic grocery bags you take home.

- Many grocery stores recycle plastic bags. Does yours? Find out this week, then, if possible, recycle any you've got at home.

So You Want to Do More

reduces CO_2
improves health
$ saves money
saves time

- Avoid purchasing two or more processed food items a week, which use up more energy in production than fresh foods and meals you prepare yourself. If you typically buy frozen green beans, for example, buy fresh instead. If you usually order takeout twice per week, forego it one time. You'll also be saving on throwaway packaging.

- Bring your lunch to work with you at least once a week. 🖋♡$

- Read labels to find out where your food is coming from. If you don't see much that's local, ask your grocer to stock more food from nearby. ♡🖋

- Plant your own garden. No yard? Try indoor pots. No green thumb? Start with something simple, such as herbs. 🖋

- Read Michael Pollan's *The Omnivore's Dilemma*. A fascinating look into state of the American food chain, Pollan's book starts with the question "What should we have for dinner?" The answer, it turns out, is rather complicated, as he follows corn from farm to feedlot, peeks into government regulations, forages for his own supper, and debates the ethics of eating animals. ♡

- Join the Slow Food movement, a cultural organization that believes that big agriculture and fast food are homogenizing foods and flavors and is founded on the principle that delicious, fresh meals and good company go together like natural peanut butter and organic jelly. Founded in 1986, today the organization is active in fifty countries and has a worldwide membership of more than eighty thousand. ♡

- Purchase extra fruit during summer and fall from your local farmers market and dry or can it. That way you'll have local produce available year-round, even when it's not in season. 🖋

- If you're like the majority of Americans who want to know if their food contains ingredients that come from cloned animals, let the government know. The Union of Concerned Scientists Action Center website, which tackles this issue and many others, is a great source for contacting policymakers. ♡

- Stay abreast of food news and issues with TreeHugger.com's food and health coverage. For delicious, planet-friendly recipes and tips, visit PlanetGreen.com.

5

Week Three

Greening Up Your Act:
Cleaning and Interior Décor

Your mission:
Save your health—and the planet—by detoxifying
the air in your house.

Picture this: You're walking down a busy city street, cars and buses chugging along, belching emissions into your path. You retreat into the safety of your own home, close the door on the polluted scene behind you, and take a deep breath of nice fresh air. Or so you think. Indoor air quality can be two to five times worse than that outside, and most people spend about 90 percent of their time indoors. In fact, the EPA lists indoor air pollution as one of the top five environmental risks to public health. But you're clean, caring, and neat. So how can your home be filled with toxins?

Of all the household chores, cleaning usually falls pretty low on the list of enjoyable tasks—who wouldn't love anything that makes the job a little easier? Enter our seduction by chemicals. Conventional cleaning products make dirt disappear fast, but unfortunately, they themselves don't just vanish. What's left behind can be bad for your health, and can flow into the municipal groundwater and waste stream when you clean the toilet or the tub. The good news is that removing a number of toxins from your home is relatively easy.

SAFETY FIRST: THE PRECAUTIONARY PRINCIPLE

There are more than eighty thousand chemical compounds approved for use by the EPA in the United States. Of these, only about a fraction have publicly available reports of evaluations for human safety. Only about 20 percent of the eighty thousand are in commercial use at any time, and federal regulations and liability issues mean that almost all new chemicals have some degree of testing or structural analysis for impacts on human health and the environment. However, these reports are interpreted by companies with financial interests in selling the chemicals and are not required for review by independent bodies. Still fewer tests have been done on how combinations of chemicals affect us, which is how we are typically exposed.

Not all chemicals are evil and not every natural ingredient is harmless. Without a body of scientific knowledge, it can be tough to pin diseases on specific chemicals. On the other hand, it is logical to correlate an increase in chemical production and societal use to an increase in many diseases.

What we do know is that hazardous substances commonly used in cleaning products, pesticides, air fresheners, and disinfectants, among other things, linger in dust and air before ending up in our bodies. According to one EPA study, every American—kids included—has at least seven hundred pollutants in his body. Rising cancer incidences since 1940 parallel our increasing use of chemicals over the same period. One half of the world's cancers occur among people in industrialized countries, and 80 percent of all cancers may be attributable to environmental influences.[1] Certain substances used in household products are also known to cause developmental problems in children; multiple chemical sensitivity syndrome; cancer; adverse effects on the reproductive, immune, and nervous systems; and a host of other illnesses. Other substances we don't know much about at all.

The U.S. government takes an innocent-until-proven-guilty approach to chemicals, and cleaning product manufacturers are required neither to disclose all ingredients nor to label their wares for anything but acute (immediate and severe) re-

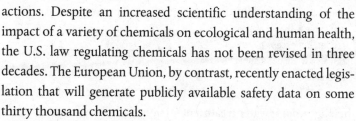

What is the use of a house if you haven't got a tolerable planet to put it on?
—Henry David Thoreau

actions. Despite an increased scientific understanding of the impact of a variety of chemicals on ecological and human health, the U.S. law regulating chemicals has not been revised in three decades. The European Union, by contrast, recently enacted legislation that will generate publicly available safety data on some thirty thousand chemicals.

Hence, at TreeHugger our motto is safety first. If an action or product is suspected of causing irreversible harm to the public—even if scientific proof is lacking—then precautions should be taken until its nonhazardous nature can be confirmed. This approach is known as the precautionary principle.

HOUSEHOLD CHEMICALS AND THE ENVIRONMENT

Along with damaging your health, household cleaners and other products can also harm the planet. Most conventional household cleaners are made from petrochemically derived chemicals; gathering raw materials, transporting, and using and disposing of these products also creates threats to the land and depletes resources. Not to mention the fact that most of them arrive on store shelves packaged in fossil fuel–generated plastic packaging.

Emissions from an individual household may seem insignificant, but multiplied by millions of residences, they can have a severe effect. In the Los Angeles region, emissions from household products are projected to exceed tailpipe emissions as the number one source of smog by 2020 by some estimates.[2] Smog, which is caused by ozone, can also be caused by hydrocarbon chemicals found in products such as aerosol deodorizes. And washing toilet bowl, surface, and other household cleaners down the drain can take its toll on water systems. Because of their age, many water treatment facilities, while well suited for removing human waste, aren't equipped to deal with such a toxic tsunami of chemicals.

Eco-myth:

Without smog, we wouldn't have beautiful sunsets.

Fact:

Sunsets appear naturally, regardless of pollution. Aerosols in the atmosphere can make a sunset appear redder, but excess pollution can also wash out a sunset's colors.

CANCER AND ASTHMA ON THE RISE

According to the Environmental Working Group, a nonprofit scientific research organization focused on health and environmental issues, the chance that a man living in the United States will develop cancer at some point during his lifetime is one in two; for women, it's one in three. But only 5 to 10 percent of all cancers are directly linked to inherited genetic factors, leading researchers to believe that environmental factors are playing a significant role.[3]

Hormone-mimicking chemicals known as endocrine-disrupting compounds, or EDCs, for example, could induce delayed cancers in the breast, testicle, or prostate gland. Other chemicals can reduce fetal growth rates and may contribute to infertility. So why do some people develop diseases while others remain healthy, even if they share similar lifestyles, product choices, and living situations? Each individual's genetic makeup is unique, and so, too, their chemical sensitivities.

Researchers also believe that more than two hundred chemicals found widely in pesticides, cosmetics, dyes, drugs, and gasoline and diesel exhaust that are known to cause breast cancer in animals may also be linked to—surprise—breast cancer in humans. A recent study by the Silent Spring Institute, a scientific research organization that focuses on the environment and women's health, looked at exposure levels to EDCs in the homes of women with breast cancer. The report—the most comprehensive of its kind to date—found an abundance of chemicals from plastics, detergents, personal care products, disinfectants, adhesives, flame retardants, and pesticides. Though the research doesn't prove a direct link to cancer, the report calls the findings "strong enough that many of us would want to reduce exposure as a precaution."[4]

A rise in asthma may also be attributable to our indoor environments, according to scientists who believe there is a link between the disease and the volatile organic compounds, or VOCs—carbon-based chemicals that form vapors at room temperature—found in things such as cleaning products and paint. About 7 percent of the U.S. population suffers from asthma, which is quickly becoming a leading childhood chronic illness. Also at fault: car exhaust, mold, tobacco smoke, and other sources of air pollution. And as global warming produces more bad-air days, asthma sufferers can expect more problems, too.

Federal law mandates that pesticides be "safe" for newborns and children but provides no explicit protection from other commercial chemicals. Many experts suspect that exposure to environmental pollutants early in life may be contributing to childhood diseases such as autism and attention deficit disorder, which are on the rise. Other health issues could surface later in life. Even in the womb, a baby can be exposed to as many as one hundred hazardous chemicals, and if the U.S. Food and Drug Administration regulated breast milk, some wouldn't be allowed on store shelves.[5] (Breast milk does, however, still provide excellent immunity and nutritional value, and generates much less waste than formula feeding.)

CONVENTIONAL HOUSEHOLD PRODUCTS

When we talk about cleaning agents, we're talking about a variety of types of ingredients. Surfactants, solvents, and synthetic dyes and fragrances are the main things that make modern cleaners both effective and dangerous. Often they are persistent—that is, they don't readily biodegrade in the environment or in our bodies—bioaccumulative, and toxic. Many have been linked to serious hormonal, reproductive, neurological, and immune disorders.

Surfactants

Surfactants, which appear in things such as dishwasher detergent and toilet bowl cleaner, include the range of chemicals that work at the surface to dissolve water and oils. Many are perfectly safe, but complex surfactants can be persistent, and some, such as nonylphenol ethoxylates, may be hormone disruptors.

Solvents

Solvents appear in oven cleaners, paint removers, furniture polishes, glass cleaners, spot removers, air fresheners—the list goes on—and are used to dissolve fats, oils, and grease and to prevent clumping. Most are highly volatile, meaning they easily become vapors, which are more readily inhaled, and are irritants to eyes, skin, and mucous membranes.

Action: Annual Savings in pounds of CO_2 • Buy an Energy Star–rated

Off-gassing

If you've ever experienced dizziness, headaches, or an asthma attack while using household products, you probably had a run-in with volatile organic compounds "off-gassing" from the product. (Carbon is the element that puts the "organic" in VOC; it isn't associated with the idea of "healthy" or "natural," as in "organic food.") Generally speaking, if you smell something—including a fragrance—you're detecting a VOC. Synthetic dyes, typically derived from petroleum or coal tars, usually contain these gases as well. VOCs, which can cause neurological and organ damage and cancer, can also off-gas from air fresheners, hairsprays, deodorizers, and carpets and furnishings, as well as many more things.

> Known to cause cancer in animals, mothball vapors can have serious short-term toxic effects. Cedar trunks and chips are a good substitute; if you do choose moth repellents, use them in trunks or other containers in ventilated areas, such as attics and detached garages.

Phthalates

Manufacturers use a class of chemicals called phthalates mainly to extend the life span of scents in their products and to render plastics flexible. They can be found in a wide array of common products, from all-purpose cleaners to cosmetics, where they are typically used to fix the perfumes. Some phthalates may cause reproductive and developmental disorders, cancer, organ damage, or childhood asthma and allergies.

Formaldehyde

In addition to being found in foam cushions, wallpaper, carpet glues, ceiling panels, insulation, and pressed-wood products such as plywood, formaldehyde can sneak into products such as deodorizers, room fresheners, bed linens (anything marked "permanent press" is suspect), and even paper towels. A very common indoor air pollutant, this compound is also a VOC. It's colorless and has a sharp odor—that strong whiff from new furniture made from pressed wood is typically formaldehyde—but is usually released so slowly in the home that it cannot be noticed.

Antibacterial Products

With our society's increasing tendency toward hyper-cleanliness, we are weakening our bodies and creating virulent strains of resistant germs and bacteria. Because they prevent our immune systems from getting a workout, antibacterial products—which aren't effective against viruses, by the way—may be contributing to asthma, allergies, and other problems. Products marketed as antibacterial are intended to kill organisms, and are thus regulated by the EPA, which classifies them as pesticides. Disinfectants and sanitizers also fall into this category.

THE GREENER THE CLEANER

"Okay," you're probably thinking. "I get it. Conventional cleaning products are the spawn of the devil. So how am I supposed to get my toilet clean?" First of all, we want to point out that people are not dropping like flies every time they mop the floor. You don't have to throw out your half-used containers of conventional Pledge and Windex right this instant. (If you do, however, be sure to dispose of them properly. Some household products and pesticides are considered hazardous materials, a good indication of their toxicity.) You can make changes incrementally by replacing conventional products with greener ones as you run out, or by making your own. It's easier than you think.

Greener household products are free of VOCs and toxic chemicals and are readily biodegradable. They use essential oils as scents, instead of synthetic fragrances, and have no fumes. They should avoid petroleum-based solvents like benzene and toluene and glycol ethers, chlorinated chemicals, phosphates, synthetic dyes, and other irritating materials. They employ plant-based surfactants. All this, and they must still effectively tackle dirt and grease.

> It's a no-brainer that products labeled as "cruelty free" or "not tested on animals" are a kinder, gentler sort. In assuming that such products are safe for human use, however, manufacturers must rely on research that was conducted on animals using the same or similar chemicals.

Though nontoxic and natural cleaners are significantly better for the environment and human health than their conventional counterparts, they aren't necessarily impact-free. The plant-based raw materials—typically corn, coconut, and soybeans—on which they rely could be grown using pesticides or genetically modified organisms, on monoculture farms, or on deforested land. Others use small amounts of petrochemicals as preservatives that prevent bacterial buildup and clumping.

As the emerging field of green chemistry gains momentum, we're beginning to see pollution prevention at the molecular level. The field focuses on redesigning chemicals to be of lower toxicity, have minimal environmental impact, and create less waste. Someday we may use household products whose every ingredient is organically grown, GMO-free, plant-based, readily biodegradable, 100 percent nontoxic, *and* that kicks dirt's butt. Until then, using existing environmentally preferred products is the best way to support this trend. As one Greenpeace report on toxic household chemicals notes, "Consumer pressure is a major incentive" in motivating companies to make changes.[6]

Better Housekeeping

It's tough to quantify indoor air quality with the naked eye, but toxic chemicals and poor ventilation are the biggest contributors to undesirable air quality. Unless you've got serious health problems, you can forego testing kits and professional monitoring, and follow these simple steps, which should significantly help you in clearing the air. Remember that all of these tips go double for the littlest TreeHuggers—our kids and pets—whose small sizes and undeveloped immune systems render them particularly susceptible to environmental pollutants, especially since they like to lick things and they play on floors and furnishings, where toxins tend to collect.

- Choose nontoxic and natural household products and reduce or eliminate the use of pesticides, especially for children's rooms, areas where pets hang out, and in the kitchen. Usher kids and pets outside while you clean.

- Open windows and ventilate, even in wintertime, especially when you are cleaning or using glues, paints, or solvents. Wear gloves and goggles and even a mask if you detect any sensitivities.

- Use a reusable microfiber cloth instead of paper towels to clean. These create less dust and less waste.

- Don't use hot water with toxic chemicals; it can cause them to off-gas more easily, releasing VOCs. Dilute cleaners with tepid water whenever possible.

- Rinse surfaces with water after you clean them, which removes toxic residues. Avoid synthetic waxes and polishes, which leave residue behind.

- Avoid spray cleaners that create fine mists, which disperse tiny particles into the air and spread around your home more readily.

Better Housekeeping (continued)

- Avoid tracking pollutants into your home by removing shoes at the door, placing a rug at each entry, and vacuuming frequently. By some estimates, about two thirds of house dust is tracked in from outdoors.

- It's a no-brainer, but if you smoke, avoid doing so indoors, or ventilate very well. Secondhand smoke is made up of more than four thousand compounds; more than forty of these are known to cause cancer and are irritants.

- To prevent soot and gases such as radon and carbon monoxide from collecting in your home, properly vent gas and wood stoves, fireplaces, furnaces, hot-water heaters, and anything else that combusts.

- Choose candles made from beeswax or soy, which are cleaner burning than petroleum-derived paraffin, and avoid wicks with a metal strand in the center. Also, scented candles, unless made with essential oils, produce more soot than unscented varieties, which may be bad for your health.

READING THE LABEL

Though manufacturers must disclose if products are acutely dangerous—that is, if a small amount can cause harm—they are not required to disclose all ingredients, nor must their formulas be third-party-tested for safety. Disinfectants and sanitizers are the exception, since they are considered pesticides. Some manufacturers market products to mislead the consumer into believing that a product is environmentally benign when it is not. This tactic, increasingly being used in many industries, is called "greenwashing."

Interpreting Labels

Here's a quick reference for understanding some of the terms we see on labels, eco-oriented or not:

- While many ingredients are "biodegradable," the term doesn't necessarily exclude materials that could take hundreds of years to decompose.

- While there's no "USDA Organic" label for cleaning products, some plant-based ingredients are labeled "certified organic." Look for a third-party certification, such as Oregon Tilth.

- "Inert ingredients" is not equivalent to "harmless ingredients." "Inert" simply refers to ingredients that are inactive in the marketed function of the product. That is, an inert ingredient in a pesticide could be harmful to you, just not to the targeted pest.

- Look for companies that mention that they disclose all of their ingredients, such as Seventh Generation. Or visit the websites of European detergent manufacturers, which must follow European law to make a data sheet with all ingredients of their products publicly available.

- "Oxy-" and "oxi-" are prefixes commonly used for eco-cleaners and detergents. These use sodium percarbonate, a powder that combines soda ash with hydrogen peroxide. Gentler on the environment than sodium perborate (another "oxygen bleach"), sodium percarbonate also functions in cooler water and is a sound alternative to chlorine bleach.

- Ingredients that start with "phth-" or "chlor-" indicate the presence of phthalates and chlorine, respectively. Anything ending in "-ene" or "-ol" is likely a petroleum-based solvent. "Phenol" in the name points to coal tar derivatives, while "glycol" indicates other petroleum-based compounds.

DIY FORMULAS

What did people do before Formula 409? They didn't live in filth—they whipped up homemade cleaners. Doing so saves money, cuts down on waste, and ensures you know exactly what you're squirting on countertops and windowsills. Do-it-yourself cleaners made from safe ingredients are healthier and readily biodegradable. Keep in mind, of course, that caution should always be used: In combination, even natural ingredients can react negatively or irritate skin and eyes.

- **Distilled white vinegar.** Our favorite common ingredient, this is an all-around disinfectant, surface and glass cleaner, degreaser, fabric softener, and more.
- **Hydrogen peroxide.** This works like bleach and also deodorizes.
- **Hot water, lemons, and sunlight.** These three will brighten and sanitize clothes and surfaces; lemons and other citrus oils will also cut through grease and smells.
- **Baking and washing soda.** These make good detergents, remove odors, and add scouring power, as does borax.
- **Soap.** Don't feel comfortable without some suds? Try castile soap.

Clean Sweep: Household Products

Don't have the time to decode the label on every green cleaning product? We've got your back. Consider this cheat sheet a quick reference guide to the top TreeHugger-approved products for making light work of dirty chores.

Ecover Floor Soap
Why we like it: Made from plant-based ingredients, Ecover's completely biodegradable floor soap leaves no chemical residue. And because it comes in concentrate form, you'll get more mileage per fluid ounce.

Seventh Generation Liquid Laundry Detergent
Why we like it: Seventh Generation's vegetable-based laundry detergents are nontoxic, biodegradable, and contain no chlorine, phosphates, artificial fragrances, or dyes. Plus, they work great.

Clean Sweep: Household Products (continued)

Method Hand Wash

Why we like it: With a dispenser that's much chicer-looking than that of your average hand soap, Method's hand wash is made from naturally derived, biodegradable ingredients. Bonus: Contains vitamin E and aloe.

Aubrey Organics Earth Aware Household Cleanser

Why we like it: This able multitasker uses an all-herbal formula that combines soap bark extract and coconut oil soap with geranium, rosemary, and sage. It's also 100 percent natural and biodegradable.

Mrs. Meyer's Clean Day Toilet Bowl Cleaner

Why we like it: Cruelty-free, septic-safe, and made with biodegradable ingredients, this toilet-bowl buster is also an aromatherapy booster.

Shaklee Dish Wash Automatic Concentrate

Why we like it: Minimalism is our thing. And this dishwashing formula knows it. Just two teaspoons of Shaklee's superconcentrated dishwashing liquid are required for an entire load, and the liquid's enzyme-activated cleaning power gets the job done without any phosphates, chlorine, or fragrances.

Biokleen Oxygen Bleach Plus

Why we like it: Whitens and brightens clothes sans phosphates, chlorine, petrochemical solvents, or other chemical nasties. Three times more concentrated than regular bleach products, Oxygen Bleach Plus contains grapefruit seed extract and natural fabric and water softeners to rinse clean without leaving residue. Plus, it's safe for colors.

AIR PURIFIERS

Whether or not air purifiers, also known as air cleaners, can help improve indoor air quality is a subject of debate. Some purifiers remove small particles from the air; others also remove specific gaseous pollutants. From inexpensive tabletop models to whole-house systems, these devices come in three varieties: mechanical filters, electronic air cleaners, and ion generators. Whether they filter pollutants or reverse the charge of particles to attract and trap them, their effectiveness depends on their efficiency and draw.

Because no air-cleaning systems "effectively remove all pollutants," according to the EPA, source control—that is, eliminating problematic materials and contaminants at their origin—is the best solution for improving indoor air quality. Ventilation is the next best choice. The EPA suggests using mechanized air cleaners only if source control and ventilation have not worked sufficiently.

The Ozone Factor

Some electric and ionic purifiers produce ozone unwittingly. Others do so purposefully to help reduce odors or render various pollutants harmless. This is a controversial method. (Ozone in the upper atmosphere filters ultraviolet radiation from the sun; in the lower atmosphere it can be harmful to the respiratory system and create smog.) While ozone from air purifiers may mask odors, it reacts too slowly with certain pollutants to remove particles from the air and forms harmful compounds with others.

While one experiment showed that ozone did indeed reduce airborne chemicals from new carpeting, it also found that the reaction produced a variety of aldehydes—highly reactive organic compounds—and that the total concentration of organic chemicals actually increased.[7] Another showed that while typical use of ozone-emitting devices doesn't reach unsafe levels, using air fresheners simultaneously in the same room could lead to unsafe exposures to formaldehyde, a particular problem for children.[8]

Using Plants to Clear the Air

If houses here on Earth have poor indoor air quality, imagine the issues a space dweller might have. In the early 1970s, scientists working on Skylab III realized they had a problem: three hundred different VOCs circulating within the space station's atmosphere. Taking a cue from life on this planet, NASA used a dozen varieties

Using Plants to Clear the Air (continued)

of plants to clear the air of compounds such as formaldehyde and benzene.

Today, researchers have identified even more plants that can do your dirty work. Easy-to-grow types such as Boston fern, areca palm, and peace lily top the list; each absorbs different pollutants, so the more you have, the better. You'll still have to pick up the mop and the citrus-soy surface cleaner, of course, but houseplants help absorb gases from the air, then store and render them harmless in their roots. Plus, soils and leaves emit water vapor, which helps keep air properly humidified, and plants give off phytochemicals that defend against mold spores and bacteria. The U.S. EPA contends that more evidence is needed to prove "that a reasonable number of houseplants [can] remove significant quantities of pollutants," but we're with NASA's space nerds on this one: Indoor gardening gets the green thumbs-up.

FURNISHINGS AND OTHER DÉCOR

Cleaning agents and other household products aren't the only things affecting indoor air quality. By some measures, furniture accounts for the largest surface area in the interior environment, rendering it a big player in indoor pollution. Pressed wood, multi-density fiberboard, and plywood are made using adhesives, and are typically chock-full of VOCs. Since they're not long-lasting materials, they're also landfill bait. Finishes, paints, subflooring, and cushion foam can also off-gas, and textiles can be saturated in toxic flame retardants. Eco-friendly furnishings and accessories exist, however, that are made from materials that have minimal impact on natural and interior environments and maximum performance value, and are either recyclable or biodegradable.

The most sustainable materials, though, are those that aren't used at all. Modern eco-designers embrace this philosophy, using a minimum of materials and energy to accomplish their goals.

Though the parameters of eco-furniture are far-reaching, a few things to look for include biological or recyclable materials; non-toxic finishes, such as linseed oil, pine resin, and beeswax; natural fillings such as cotton, wool, and down, as opposed to foam; and reclaimed or salvaged materials.

Durability itself is an eco-friendly quality; it translates into fewer purchases over the long haul. On a budget? Securing secondhand furniture can yield sturdier finds than buying new at the same price. Likewise, selling or donating keeps landfill detritus in check. Plus, used goods are often bought and sold locally, reducing energy consumed for transport. The same thing goes for locally made items and locally sourced materials.

> If every household in the United States replaced just one package of virgin-fiber napkins with a 100 percent recycled pack, we could save
>
> # 1 million
>
> trees from being cut down.

Design for disassembly—another quality we like—allows for products to be broken down into their component parts for reuse at the end of their life. It also indicates that items can be easily fixed, as opposed to needlessly tossed. "Monstrous hybrids"—things that contain multiple types of materials that are fused together—on the other hand, achieve the opposite effect.

Sustainable Furniture Materials

Aside from air quality issues, home furnishings and accessories can have a multitude of additional environmental impacts. The biggest issue with furniture and anything else made from wood is irresponsible logging—clear-cutting or harvesting from endangered or protected regions—which can lead to deforestation, a precursor to global warming. Products made from trees that grow quickly, such as birch, beech, and maple, for instance, are generally considered preferable; tropical hardwoods such as teak or mahogany that aren't certified sustainable might raise a red flag.

Though it can be tough to determine if wood comes from sketchy sources, manufacturers are increasingly demanding materials from well-managed forests, and typically tout it when they do. Third-party certifications such as that from the Forest Stewardship Council, indicate that lumber has been harvested from well-managed, self-sustaining forests.

Fast-growing and versatile, grasses are also quickly becoming choice materials. Bamboo, some species of which can grow up to four feet in one day, is incredibly strong and can be fashioned into furnishings, flooring, flatware, accessories, and structural material. Kirei and Wheat Board, made from sorghum and straw, respectively, and Plyboo, made from bamboo and low-emitting adhesives, are ecologically sound and attractive alternatives to conventional plywood.

Though metals and plastics represent high levels of embodied energy—that is, the energy required from extraction to finished product—both are theoretically recyclable and can also be considered eco-friendly under the right circumstances. Furniture made from recycled materials saves energy and natural resources. And trust us, "recycled" no longer means "rustic" (read: ugly)—today's hip reborn materials have serious aesthetic appeal.

Wallpaper

Covered with sprawling roses, bedecked with toile, or stamped with mod graphic patterns, most wallpaper isn't made from paper at all but from vinyl, and the adhesives used to tack it up are loaded with VOCs. Residential vinyl-free wall coverings do exist, and low-emitting adhesives are becoming a bit easier to find, but to date, eco-improvements in wall coverings have been limited somewhat to commercial applications. Luckily, interior paints are picking up the slack.

Paint

Paint can be a major indoor environmental hazard and freshly painted homes can have VOC exposures up to a thousand times greater than normal. Regular latex and water-based paints off-gas less than their oil-based counterparts, but now a broad spectrum of options is available—ranging from high-end paints based on Le Corbusier's color theories to more accessible offerings from brands such as Sherwin-Williams and Farrow and Ball. Labeled low-, no-, or zero-VOC, they all release some gases, but at greatly reduced rates. Even "natural" paints, which are made sans synthetic chemicals, off-gas somewhat. Milk paints emit no VOCs, but can be challenging to work with and aren't as long-lasting.

Textiles

Textiles made from natural materials such as cotton require energy, water, and dyes—which can contain heavy metals and other harsh chemicals—to produce. Some textiles, such as those made from hemp or silk, are by their very nature more ecologically sound. Bamboo and bamboo-blend fabrics can be terrifically soft. Hemp—which is beginning to lose its hippie reputation (though it's still illegal to farm in the United States)—is an incredibly resilient fiber that's fast-growing, naturally pest- and mildew-resistant, and requires little fertilizer.

Sustainable fabrics use a combination of renewable resources, low-energy and clean production processes, safe chemicals and dyes, and biodegradable or recyclable materials. While it still takes some effort to source sustainable fabrics for the home, they are becoming increasingly available. The commercial interiors industry already has a wide array of choices, which bodes well for residential consumers, who will find organic cotton increasingly available. Other fabrics—such Ingeo, which is made from corn (although the use of genetically modified crops is a subject of debate), and several made by Designtex—can biodegrade or be endlessly recycled.

Flame-retardant Fabrics. Many upholstery and drapery fabrics—as well as mattresses, bedding, kids' pajamas, and electronics—are treated with flame retardants, which can end up in dust. These, of course, serve a very important purpose—slowing the spread of fire—but many environmentalists consider them suspect. Several flame retardants have been banned due to their persistent nature and health-safety concerns; a few states have also banned others that are still allowed under federal regulations. Encouragingly, visionary upholstery manufacturer Climatex has developed a Cradle to Cradle–certified fabric made from wool and the cellulose from beech wood that is both compostable and flame-retardant.

Stain-resistant Fabrics. To allow potential stains to slip off your couches and carpets, chemists use perfluorochemicals, or PFCs, which are also found in clothing, food packaging, and nonstick pans. Sure, they make life easier in some ways, but PFCs can release persistent toxins into the environment and have been linked to a host of health disorders. You don't have to toss out every Teflon-coated pan you own—research shows that exposure from cookware is likely very limited. (Still, a few precautions can't hurt: Ventilate when cooking, don't place empty pans over high heat, and get rid of pans that have started to flake.) We do know that PFCs are being found in the environment and our bodies. While the majority are released during manufacture, small quantities are released as the products containing them are used and break down in the home.

Flooring

While a few industry giants such as Interface, Shaw Industries, and Mohawk Industries, for example, have made impressive strides in greening their products and businesses, many carpets still contain glues and adhesives or PVC backing, which can off-gas into your home. If buying and installing carpeting that you suspect contains VOCs—"new carpet smell" is a tip-off—let it air out thoroughly before installation, or ask the retailer to do so for you.

Choosing carpets and rugs made from all-natural fibers or recyclable nylon can be a healthier option. Hardwood floors, another good choice, don't trap dust, pollutants, or allergens and are easy to clean. Other environmentally preferable flooring choices include linoleum, which is made from linseed oil; recycled/content tile; and cork, which is naturally antimicrobial, soft underfoot, and renewable, as it comes from bark that is harvested from a tree roughly every nine years, with no harm to the tree.

Kid Stuff: Tips to Live By

Kids are especially susceptible to environmental toxins, so it makes sense to take extra precautions to keep them safe. You can't protect your little ones from everything, of course, but you can minimize their exposure to harm by following a few simple tips that will help you create a greener home, safer meals, and a better world for them to grow up in.

- Eat right. Pregnant women, nursing moms, and young children should avoid fish such as tuna, tilefish, and shark, which are high in mercury, and other seafood with high concentrations of bioaccumulative toxins. Eat fish lower on the food chain instead.

- Skip the beauty salon. Many pregnant and nursing women choose to give up hair dye and nail polish, which give off fumes and contain other toxic ingredients that could enter their bloodstream. Though these beauty treatments have not been proven to be harmful, many health experts warn that they could be. Avoid bringing babies and young children into beauty salons, which exposes them to unhealthful fumes.

- Renovate with care. Major home renovations are a big no-no during pregnancy or infancy since indoor air quality can suffer significantly. If you must renovate, seal off that section of the house as well as possible. Hire someone to put up wallpaper and paint. Ventilate as much as possible, and wait a few days before allowing a child of any age to sleep in a redone room.

- Decorate appropriately. Babies and children are especially susceptible to VOCs and other toxins. Take care to decorate nurseries and kids' rooms with low-VOC paints and furnishings; choose carpeting that doesn't use adhesives or PVC backing; and shop for phthalate- and formaldehyde-free mattresses made from natural materials.

- Feed with milk and formula wisely. Breastfeeding generates the least amount of waste and builds baby's immunity. Store breast milk in reusable containers, rather than one-time-use bags. Bottles made from polyethylene, polypropylene, or tempered glass are better options than polycarbonate. Powdered formula has a smaller transportation footprint than liquid since it is lighter in weight. Plus, cans used to store some liquid formula have been found to leach bisphenol-A, a hormone disruptor, into the formula, sometimes at levels two hundred times greater than FDA-regulated levels.

- Choose healthy foods. Organic food for baby means exposing him to fewer toxins.

- **Keep clean.** Greener cleaning products are a must with kids in the house.
- **Do your diaper duty.** Disposable diapers are used by just 5 percent of the population, but are the third largest source of landfill waste. One better option includes flushable and compostable gDiapers. Cloth diapers cut down on waste but can use twice as much energy and four times as much water as disposables. The "elimination communication" technique solves the diaper-use issue altogether, requiring parents to pay attention to baby's body signals. Seventh Generation diapers are disposable but chlorine-free and use nontoxic absorbant polymers.
- **Make bath time safer.** Organic grooming products ensure that soaps, lotions, and oils are gentler on baby's skin and the planet. Avoid perfumes and fragrances—this goes for adults, too.
- **Dress for the future.** While the fibers your kids' clothes are made from won't necessarily directly affect their health, buying clothing and accessories made from greener materials such as organic cotton and hemp help support a more sustainable future. Sleepwear can be coated with flame retardants; these serve an important purpose but can also contain harmful toxins. Also, keep kids clear of freshly dry-cleaned clothing.
- **Grow greener grass.** Using organic lawn care ensures that kids won't be exposed to toxic fertilizers, weed killers, and other chemicals while playing in the yard.
- **Get involved.** Concerned about a school's indoor air quality, lunchtime meals, or playground surroundings? What about diesel bus fumes and parents in idling cars waiting to pick up their little ones? Speak up, act out, or join the PTA. Nobody gets heard like an educated, concerned parent.

Toy Story

Aside from lead paint, the main substance of concern found in toys is PVC, or polyvinyl chloride—a very common type of vinyl that releases toxins into the environment throughout its life cycle. It can be found in everything from beach toys and teethers to dolls and rubber ducks. Many toys made with PVC also contain phthalates, chemical plasticizers used to make the material softer and more flexible. Though the long-term effects of phthalates on youngsters haven't been fully established, in 1999 the European Union enacted emergency legislation to ban six phthalates from toys designed to be put in the mouth for children under three.

Fortunately, many manufacturers are taking the initiative to reduce or eliminate PVC and other hazardous materials from their wares. Playmobil, for example, eliminated PVC from its children's products more than twenty years ago, and all LEGO toys for kids under three have been PVC- and phthalate-free since 1999. Choosing toys known to be PVC-free or calling the manufacturer is the best defense, but even the greenest parents know that the pressures and speed of modern life can cause convenience to trump ideals.[9]

Switching to ecologically friendly cleaners is relatively easy. Eliminating plastics from your kid's life is damn near impossible. We try not to get our organic cotton panties in a bundle over every teething toy, building block, and stuffed bear. Instead, we try to find the balance among getting educated, being protective, and not freaking out. In a world filled with flashing whirligigs, exclusive preschools, and scheduled playdates, we offer here a few tips for some good old-fashioned, phthalate-free fun:

- Toys made from untreated wood are sweet, durable, and old-timey. Toys made of nontoxic coatings and paints and nontoxic craft supplies are also winners.

- Cutting back on playthings that require batteries may not only save your sanity (Does "If that pseudo cell phone rings one more time, I'm going to lose it!" sound familiar?), but also reduce the chances that batteries will find their way into tiny mouths—or the landfill, for that matter.

Toy Story (continued)

- Since kids outgrow toys quickly, trading with family members or buying secondhand items cuts down on manufacturing waste and saves you some loot. Or join or start a toy lending library.

- Because they aren't grown with pesticides and fertilizers, we like kid stuff made with organic fibers, such as cotton and wool, and natural dyes.

Pet Projects

We love our pets like family, and judging by the approximately $38-billion-a-year U.S. pet goods market, you do, too. We've said it before and we'll say it again: You can't shop your way green, and minimizing your footprint means purchasing fewer non-necessities for Fido and Fluffy, too. (Come on, now. Does Rover really need another biker jacket?) Still, pets do need some stuff. Luckily, Earth-friendly options are available.

Organic pet food, natural grooming products, and toys made from sustainable fibers are increasingly easy to find, and you can probably guess that using a pooper scooper or biodegradable bags (we like BioBags, made from GMO-free corn) is a better choice than enshrining doggie doo in everlasting plastic grocery bags that could instead be recycled. What about Miss Kitty? We've heard of cats using the toilet, which sounds handy and eco-correct. But if your cat isn't quite so civilized, trade in the clay litter—which is obtained through strip-mining and contains harmful additives that end up in dust—for something more eco-friendly, such as litter made from wheat or corn. To keep little paws comfortable come winter, try a salt-free de-icer such as Safe Paw; it's healthier for children, and the planet, too.

Q & A

Jeffrey Hollender of Seventh Generation on Disregarding Your Mother's Traditions, Not Achieving Perfection, and Looking to the Future

What are the most important things people should know about the chemicals in conventional household cleaning products? The first two things I recommend that people transition away from are products that have hypochlorite or chlorine in them and, second, products that are disinfectants, which are basically pesticides used to kill germs inside your home.

Yet millions of consumers think that bleach smells "clean." When my mother was growing up, she thought smoking cigarettes was a perfectly fine thing to do. Many historically accepted practices, products, and ingredients are now known to be extremely dangerous. Household chemicals may not be much different than cigarettes—many are carcinogenic, likely to disrupt our hormonal systems, and likely to have adverse effects on children. At some point in the future—as is already the case in Europe—those chemicals will no longer be allowed. Right now, individuals have to educate themselves. We cannot rely on what our mothers used because they didn't have access to the information we have today.

Any further advice for greening the home? Purchase recycled paper products. It's a relatively easy thing to do, whether it's changing out the copier paper or tissue products. One of the biggest challenges we face as a society is great uncertainty about whether or not individuals can make a difference. But it's not about waiting for the govern-

Q & A (continued)

ment or business to make changes. It's essential that we as individuals do what we can do every day.

Your book *Naturally Clean* contains several recipes for DIY cleaners. Why would someone trying to sell cleaning products give out such information? While it's true that Seventh Generation is in business to sell products, that's not what gets me up every day. I think what motivates everybody at this company is the opportunity we have to have a positive effect on the world. For people who want to make their own cleaning products, we absolutely encourage that. From an environmental perspective, it has the least impact—still, most people aren't going to take the time to do it.

What changes in consumer behavior have you see since you joined Seventh Generation in the late 1980s? There have been more changes in the past year than in the prior nineteen combined. We're learning every day that health and environmental issues are interconnected and ultimately inseparable. The effects of global warming on health will also be profound. As the climate changes, the asthma and allergy epidemic that we're already facing is going to accelerate. This connection is motivating consumer behavior.

What are the most difficult aspects of producing and selling green household products? None of Seventh Generation's products is perfect. Every single one has attributes we want to improve. It's been a very, very difficult challenge to solve problems such as finding better preservatives that in many cases are not even on the mind of customers, yet can cost a lot of money to correct.

Q & A (continued)

The name of your company is based on a law from the Iroquois Confederacy: "In our every deliberation, we must consider the impact of our decisions on the next seven generations." What does that mean to you? The name is daunting, but also inspiring. It's an aspiration and a set of beliefs and ideals that I think everyone involved with this company feels tremendously proud of. The reality is that we fall short of doing what our name suggests most of the time. But it means we should look out into the future as far as we can every time we make a decision.

Jeffrey Hollender is president and CEO of Seventh Generation and the author of several books on sustainability and corporate social responsibility.

Save the Planet in Thirty Minutes or Less

reduces CO$_2$

improves health

$ saves money

saves time

- Read and understand the label of one conventional household cleanser or other product you own. Are the ingredients even listed? ♡

- Switch to at least one Earth-friendly household cleaner this week. (Note if the ingredients are listed on the label.) Replace either the cleaner you use the most or the product you're closest to running out of. ♡

- Crack the windows when you clean this week. ♡

- Fill a reusable spray bottle with one part vinegar to five parts water. Use it for minor cleanups, such as wiping down the countertops after making dinner. ♡ $ 🔋

- Upon returning home a few times this week, take the time to notice how your home really smells. Is it damp? Can you smell the carpet? Do the cleaning products linger? Scan your home for potential sources of contaminants and indoor air pollutants. Could that new particleboard bookcase be the source of your recent headaches? Is that old carpet making you sneeze? Typical symptoms of poor indoor air quality include headaches, fatigue, itching or burning eyes, skin irritation, nasal congestion, irritated nose or throat, and nausea. ♡

So You Want to Do More

reduces CO$_2$

♡ improves health

$ saves money

saves time

- If you have kids or pets, take special care to rethink the products you're using in areas where they spend time. ♡

- Room by room, do a mental audit of the cleaning procedures at your house. Which products do you use? Could you cut back on the number of harsh chemicals you use regularly? Would more frequent maintenance reduce the need for strong chemicals? Do you equate the smell of bleach with a clean house? Pay attention to how you feel and what you smell while cleaning. As you switch to greener products, notice if your perspective changes. ♡

- So you've already switched to greener cleaners. Congrats. Next step: Whip up a few homemade concoctions. You'll be amazed at how simple it is to clean with the most basic ingredients. ♡ $ ⊙

- Buy limited quantities of toxic products that you do need. The less you buy, the less you'll be exposed to and the less you'll waste. ♡ $

- Gases can leak even from closed containers. Transfer rarely used cleaners, paint cans, and other chemical products to parts of the house you use infrequently, such as the basement, but that are adequately ventilated. If you're ready to get rid of them all together, dispose of them safely. ♡

- If you're planning to remodel, donate old cabinets and appliances to organizations such as Habitat for Humanity, or find other places to recycle them.

- Make a commitment to avoid cheap furnishings. Shop for long-lasting pieces made from sustainable materials, or for second-hand wares.

- Find out which local hardware stores sell ecologically friendly paints. When you next project comes up, you'll know where to go.

- To stay up-to-date on the latest green home fashions and trends, check out design and architecture news at TreeHugger.com. Find DIY recipes for homemade cleaners in the home and garden section of PlanetGreen.com.

6

Week Four

Traveling Light:
Transportation

Your mission:
Get around without all the global warming baggage.

Since 1960, the number of miles Americans travel each year has more than tripled. Meanwhile, fuel economy of passenger vehicles has decreased.[1] Unsurprisingly, these trends have made conditions ripe for increased greenhouse gas emissions from personal travel; between 1990 and 2003, greenhouse emissions from transportation went up 24 percent.[2] Stagnant fuel economy regulations have compounded the issue, and will continue to do so if left unchecked. To reverse the trend, we need to focus on energy efficiency improvements, low-carbon alternative fuels, and increasing the efficacy of existing transportation systems. But above all, we need to reduce motorized transportation activity altogether. Building communities and behavioral trends that encourage walk-

> You must be the change you wish to see in the world.
>
> —Mahatma Gandhi

ing, biking, taking mass transit, telecommuting, and carpooling also have a large role to play in reducing pollution.

DESIGNING CAR-FREE CITIES

In 2006, for the first time ever, more than half of the world's population was living in an urban setting. As cities grow more congested with people, they also grow more congested with cars. But cities can be transformed to provide greater accessibility to workplaces, shopping, entertainment, and so on, without the use of automotive transit. Cities should be designed for people, after all, not machines. Doing so benefits the environment, public health, and the economy.

Several U.S. cities are among the world's worst when it comes to CO_2 emissions. Why? First, public transportation isn't serving us the way we need it to, and so we turn to personal vehicles. Second, because our land-use is so spread out—our suburbs sprawl; and we build enormous highways instead of building communities close to one another—we drive, a lot.

The evolution of modern cities is tied to advances in transportation. Initially we built ports and train stations for shipping and commerce and population grew around that. But it was cars and cheap oil that created the next big boom. Today, our cities are choked with motorized personal vehicles. Communities where people can walk, bike, or take public transportation to work, shop, and be entertained are healthier and more desirable places to live, however.[3]

BUILDING BETTER 'BURBS

Subsidies for highways have made them cost-efficient to build, ripening the conditions for sprawl to creep ever farther along. Since almost everyone can afford a car and gas is still relatively cheap, inefficient development patterns have cropped up nationwide—enormous strip malls, insensible bedroom communities, towns with enormous McMansions but no downtowns. We're paving over our suburban and rural spaces to make way for low-density, automobile-dependent development.

This type of growth is largely unsustainable, partly because these developments use more resources than they should, but also because they lack soul: Without vital economic and community centers, places like these never become communities. Our commutes have gotten longer, leaving us less time to spend with the kids, enjoy our hobbies, exercise, and relax. Ultimately, quality of life goes down. Meanwhile, we're increasing air and water pollution and converting our countryside into blacktop.

> In the United States,
>
> **two thirds**
>
> of all oil consumed goes toward powering vehicles, according to the U.S. Department of Energy.

At the same time, our population and our wealth are increasing. Growth is inevitable. To expand intelligently, we need to consider our existing infrastructures and our dwindling resources. By directing growth to communities where people already live and work ("infill" development), we can limit the number of impervious surfaces covering the landscape, reduce per capita car travel, make existing communities more attractive, and discourage resource-intensive new infrastructure from being built. This often means, for example, building up, not out; redeveloping older suburbs and existing structures; eliminating minimum lot sizes; and encouraging multifamily housing. Doing the opposite—encouraging low-density development and building from the ground up ("greenfield" development)—ups car travel and strains resources.

More compact development also breeds better public transport. Six to eight houses per acre can support a bus stop, according to the Smart Growth Network and International City/County Management Association. Compared with an individual traveling alone in a car, a transit bus carrying forty passengers requires one sixth of the energy per person to operate. That same bus also replaces traffic-causing cars that would stretch for six city blocks. Fifteen to twenty dwellings per acre can support rail transit. One full six-car train is equivalent to a line of moving cars going twenty-five miles per hour stretching ninety-five city blocks.[4] When you think of it this way, the commuter's choice is a no-brainer. Yet 77 percent of all workers in the United States drive alone to and from work.[5]

The Bicycle: King of Vehicles

We've cruised down Barcelona's Las Ramblas, clattered over the Arno river in Florence, and cut across New York's Central Park, coffee in hand, in record time. Yep, that's us squealing down L.A.'s Skid Row, swerving through downtown Jerusalem, and steering clear of pigs in Laos. Are we riding solar-powered scooters? Showing off in sporty electric roadsters? Rolling with the pious Prius brigade? Nope. We're mounted on the king of vehicles: our bicycles.

Down country lanes, designated paths, and busy city streets, we just can't say it enough: We love our bikes. Sure, walking is great, too, but as far as personal modes of transport go, bicycles come out on top for alleviating congestion, lowering air pollution, and keeping us in shape. Plus, who's ever needed a bike loan?

Two out of three

TreeHugger staff members don't own a car.

Absurdly, few of our countrymen agree: just 1 percent of all trips in the United States are made by bike.[6] That could be a dangerous example for the rest of the world. For China's rising middle class, for example, the car has emerged as a status symbol, while the bike is seen as the calling card of the poor. In recent years, the country's domestic market for bicycles has plummeted from 40 million to less than 25 million per year, due to

The Bicycle: King of Vehicles (continued)

government policies promoting automobiles. If emerging nations are to be allowed to develop fairly, we must find a way to help China's citizens and others both advance their purchasing power *and* keep their subsequent rising CO_2 emissions in check.

But even as developing nations embrace the car, the modern West is returning to the bike. Police officers and messengers are increasingly using two wheels to get around more quickly. In Copenhagen, City Bikes are free for the taking—and the dropping off—every few blocks. Parisians can now access public wheels every nine hundred feet. The government of the Netherlands even has a Bicycle Master Plan that includes laying bike lanes nationwide and giving cyclists traffic privileges. Meanwhile, bike technology itself is burgeoning: Well-designed folding bikes such as the one made by Strida make it easy for urbanites to carry their wheels on the subway and stash them in small living quarters. Bike frames built from bamboo—from Bleijh and Design Amsterdam's funky, streamlined Sandwich bike to Calfree's high-end racers, which offer better vibration dampening than carbon fiber—make pedaling an even more sustainable endeavor.

We envision a world where transit systems and bicycles interact seamlessly. Where designated street lanes lead to train stations with non-motorized vehicle parking. Where companies provide showers for employees who pedal-commute. Where bikes are as respected as cars, the streets are safe, and the skies are clean.

As we swing our velocipedes into gear from Black Rock City to Berlin, we think about fuel and energy savings, about greenhouse gases and good health. Sometimes we wonder why people pay fees to drive to gyms to ride bikes that don't move. But then sometimes we just roll along, enjoying the breeze in our hair and the miles behind us, since, after all, we just love our darn bikes.

THE CASE FOR PUBLIC TRANSIT

Transportation provides the perfect nodules on which to grow neighborhoods. Bus, subway, and train stops are perfect places to concentrate office and apartment buildings, shops, and entertain-

> A bicycle is a marvel of engineering efficiency, one where an investment in 28 pounds of metal and rubber boosts the efficiency of individual mobility by a factor of three. On my bike I estimate that I get easily 7 miles per potato. An automobile, which requires one to two tons of material to transport even one person, is extraordinarily inefficient in comparison.
>
> —Earth Policy Institute founder Lester Brown, in *Plan B 2.0: Rescuing a Planet Under Stress and a Civilization in Trouble*

ment centers such as movie theaters and concert halls. Linking transportation systems—rail lines, bus lines, and bike and walking paths—is the next step in creating extended, car-free mobility. If we want people to get out of their automobiles, it's essential to create scenarios where it's easy for, and indeed, benefits them to do so.

Take the example of Portland, Oregon. In the 1960s, the city, like many others, was threatened with the loss of residents, businesses, and capital. Suburban housing developments, shopping areas, and business parks were draining vitality from the city center. An emphasis on multiple modes of transit and intelligent land use helped turn the situation around.

Today, Portland boasts an impressive regional transit system of buses and a light-rail that links suburbs not only to the city but also to one another. Eliminating a freeway along the Willamette River made way for a public park. With the addition of hundreds of miles of bike trails and lanes, including one that goes to the airport, cycling traffic has gone up 257 percent during the past decade. Quashing more than 62 million car trips annually, these practices have helped Portland sidestep a predicted 40 percent increase in traffic congestion, and have brought greenhouse gas emissions down to pre-1990 levels.[7]

Other cities have found similar ways to get drivers out from behind the wheel. When London imposed a five-pound charge on all motorists driving into the city's center, traffic and air and noise pollution immediately plummeted. Despite fears of a consequent downturn in the economy, 65 percent of businesses reported no financial effect, and some retailers witnessed positive sales due to increased pedestrian traffic. Singapore, Oslo, and other cities use similar strategies to their benefit.

CASE STUDY: Curitiba, Brazil

In the 1960s, when the population of Curitiba, Brazil, began to approach half a million, some residents worried that rapid growth would drastically alter the character of the place. City planners responded with strict ideas for controlling urban sprawl, reducing traffic, and preserving the city's historic sector. The result has created a pilgrimage destination for fascinated urbanists who consider Curitiba one of the best examples of smart growth today.

Pivotal to the entire operation was improving the existing bus system by designing an express route with dedicated bus lanes. This eventually evolved into an expansive surface system rivaling sophisticated subway systems—and at much lower cost. With less auto congestion, the city has replaced several downtown streets with pedestrian malls and shopping centers, reduced polluting emissions, and helped Curitibanos pocket the money they otherwise would spend on 7 million gallons of fuel each year.

Today, despite a high rate of auto ownership, the majority of residents use public transit for routine travel—two thirds of all trips in the city are by bus.[8] Though the population has doubled since 1974, car traffic has declined by a remarkable 30 percent.[9] Because passengers can board and exit buses rapidly and buses are frequent, the system meets riders' needs—which is precisely why they use it. Because of smart planning and good public transportation, residents of Curitiba—who now number 1.6 million—have easy access to their jobs, homes, recreation, and thus a higher quality of life.[10]

The Problem with Driving

Better transit systems and civic development and behavioral changes toward the modes of transport we choose are essential elements for creating a greener future. But change will not happen overnight. The fact is that most people drive cars. There's no such thing as a truly ecologically friendly automobile, but there are greener choices we can make, from driving more efficiently to opting for a hybrid car. With global warming looming, it's high time to start implementing them.

Light-duty vehicles—that's passenger cars, vans, minivans, sport-utility vehicles, and pickup trucks—account for 62 percent of all greenhouse gas emissions caused by transportation. Passenger cars alone are responsible for 35 percent of that, the largest contributor of any vehicle category, above heavy-duty vehicles (18 percent) and aircraft (9 percent).[11]

For each gallon of gasoline a car burns, it releases about 20 pounds of carbon dioxide, so cutting your driving by just a few miles each day could literally negate tons of CO_2 from entering the air each year. (For diesel-fueled cars, that number is about 22 pounds per gallon.) Walking, biking, and mass transit are the best choices for reducing the number of miles you drive in the car, and making a voluntary switch to these modes of transport whenever possible is an imperative part of reducing your personal—and our collective—footprint. Of course, existing infrastructure means that these options aren't always feasible. If you drive a car, making it as fuel-efficient as possible is an essential next step.

Voluntarily improving fuel efficiency could go a long way. Unfortunately in the United States, we have never seen substantial reduction achieved by this method. (Don't let that stop *you* from doing the right thing.) Our increasing wealth has bumped up our vehicle ownership as decreasing household size and low-density growth has led to steadily declining vehicle occupancy rates. Travel by all modes of public transit accounts for only about 1 percent of our total passenger miles. Put another way, more people are traveling more miles in more cars, meaning more fuel is being used—and more emissions generated—per person.

For these reasons, fuel efficiency improvements, especially of cars and light trucks, are critical to reducing CO_2 emissions from transportation over the next thirty years. Several policies can contribute to our realizing this potential, including fuel economy standards. Without significant changes in U.S. energy policies, we can predict continued growth in transportation petroleum use, oil imports, and greenhouse gas emissions. Over the next decade, emissions from transportation are forecast to increase by almost 50 percent.[12] Technologically, it is already possible for manufacturers to build much more fuel efficient cars. But without government regulation and consumer demand, they have little motivation to do so.

> Most Americans think they have more control over reducing the amount of energy used in their homes than the fuel in their vehicles.

Whether you drive a Prius or a Suburban, you can start eking out extra miles from your car by following these basics:

- Keep your speed steady, and avoid sudden braking or accelerating. If your car has a manual transmission, shift into a higher gear as soon as possible. Also, driving 65 miles per hour instead of 75 increases fuel efficiency by up to 15 percent.
- Keeping tires properly inflated means less surface area comes in contact with the road and thus creates less drag. Environmental Defense reports that 32 million U.S. vehicles ride on at least two underinflated tires, wasting 500 million gallons of gas each year. Using the air conditioner also decreases fuel efficiency.
- For late model cars, idling for more than ten seconds is actu-

> If cars were required to get forty miles per gallon, it would cut global warming pollution by 600 million tons each year and save consumers more than
>
> # $45 billion
>
> annually at the pump.

ally less efficient than turning your car off and then back on, according to Environmental Defense. That doesn't mean you need to shut down at every traffic light. (Though in Japan, where conservation is practically an art form, some people do; others have cars with technology that automatically shuts off the engine when in idle mode.) But making a habit of turning off the ignition while waiting to pick up the kids at school or while in line at the bank is definitely a good idea.

- Packing light can further increase fuel efficiency. Hauling an extra hundred pounds in your vehicle reduces fuel economy by up to 2 percent. Placing luggage inside the vehicle rather than on the roof minimizes drag and increases mileage.

BUILDING BETTER PERSONAL VEHICLES

Before 2030, advanced diesel engines, gasoline or diesel hybrid vehicles, and hydrogen-powered fuel-cell vehicles could likely up fuel economy by 50 to 100 percent and cut emissions, too.[13] But automotive technology already exists that could help us move beyond our current state of vehicular disrepair. Various types of electric-assisted vehicles have become so popular over the past few years that they're hardly even considered cutting-edge anymore.

The three main types of electric vehicles are those that run strictly on batteries (EVs), hybrid gasoline-electric vehicles (hybrids), and fuel-cell vehicles (FCVs). They are labeled as low-emission vehicles (LEVs), ultra-low-emission vehicles (ULEVs), super-low-emission vehicles (SULEVs), partial-zero-emission vehicles (PZEVs), and zero-emission vehicles (ZEVs). Some traditional internal combustion engine cars earn these ratings, too—Subaru's Legacy, Outback, and Forester, for example, run on gas but are PZEV-rated and have 90 percent cleaner emissions than the average new car and lower emissions than some hybrids.

While all of these technologies are promising, none is a cure for our environmental problems. The greener route would be a trend toward smaller and much lighter cars that use less fuel to begin with and, of course, to use fewer motorized vehicles altogether. And consumer demand for better products could create a significant shift in the market, making vehicles such as the ones listed below worth mentioning.

Electric Vehicles Electric vehicles running explicitly on batteries generate zero emissions while on the go; associated emissions are those released during the generation of electricity (unless the power comes from carbon-free renewables). This is why you sometimes hear that EVs "get cleaner every year"—because each year the electrical grid adds more sources of lower-carbon energy. In general, EVs are considered 90 percent cleaner than conventional gasoline-powered cars.[14] Historically, the biggest roadblock in developing EVs has been battery technology. This is constantly improving, however, and some electric vehicle prototypes are showing ranges of 80 to 200 miles per charge while traveling at highway speeds.

The most popular electric vehicle of all time, the EV1, was manufactured by General Motors in the 1990s. But in 2004, GM mysteriously announced that it would not renew leases on any of the cars. Since then, EV1s have disappeared from roads altogether. Today, electric vehicles such as the high-end Tesla Roadster are creeping toward market availability. Capable of doing zero to sixty in four seconds and traveling two hundred miles on a single charge, the Roadster, priced at about a hundred thousand dollars, is the sexed-up face of the next generation of EVs.

A few scooters and three-wheeled vehicles, known as NEVs, or "neighborhood electric vehicles," are both street-legal and for sale now, but with top speeds of just thirty miles per hour and farthest ranges of only fifty miles per charge, their applications are somewhat limited. The Vectrix electric scooter is a promising available option that goes up to sixty miles per hour, with a range of sixty-eight miles, and plugs into any standard outlet to recharge.

Hybrids. Hybrid gas-electric vehicles have probably generated the most attention—and the most sales—of all alternative passenger vehicles on the road today. Just under a million hybrids are traversing America's roads—a fair number, but still only a fraction of a percent of all cars on the road today.[10] Hybrids such as the Toyota Prius use a battery-powered electric motor to assist a conventional gasoline-powered internal combustion engine and generate fewer tailpipe emissions than their purely gasoline- or diesel-fueled counterparts. The average hybrid vehicle gets about thirty-four miles per gallon in the city and thirty-three on the highway—city averages are higher since the electric motor can be used more often in stop-and-go and slow-speed situations. Less-efficient hybrids get as little as eighteen miles per gallon.

> **Fifty-seven percent**
> of Prius owners say the main reason they bought the car was because "it makes a statement about me."

But just because a car is a hybrid doesn't mean it's the greener decision. If a Toyota Yaris, which has a conventional internal combustion engine but gets almost forty miles to the gallon, suits your needs but you go for a hybrid with worse mileage, you're not doing the planet any favors.

Most hybrids today use nickel-metal batteries, and while no one knows exactly how long they'll last, since they are still relatively new, companies such as Toyota and Honda expect theirs to last the life of the vehicle, and warrantee them accordingly. To date, few hybrids have had major issues with their batteries, and most batteries are recyclable. The bigger issue with hybrid batteries is their capacity to store power in a small package; manufacturers are furiously working on lithium-ion batteries that will improve these circumstances.

Plug-in Hybrid-Electric Vehicles. Most hybrids on the market today use regenerative braking—that is, during braking and deceleration, energy that would normally be lost as heat is instead captured as electrical energy—to keep their batteries charged. Plug-in hybrid-electric vehicles (PHEVs), however, recharge directly from the grid. The benefit? These vehicles have a range of twenty to sixty miles before they resort to gas for power.

With PHEVs, the nearly 80 percent of people who travel approximately forty miles or less daily could make their trips using little to no gasoline.[16] And since these vehicles generally recharge at night using excess power from the grid during nonpeak times, it's possible that tens of millions of these vehicles could be charged without the need to build extra power capacity.[17] Though no plug-in cars are on the market yet, GM's Saturn VUE and Chevy Volt could land in showrooms in 2009 and 2010, respectively.

It's true that the real emissions of PHEVs depend on the source of electricity in the area where they're charged. That is, a plug-in hybrid car recharged with coal-generated electricity will have higher overall emissions than one using hydropower, for example.

One study even shows that a PHEV powered from coal-generated electricity has lifetime CO_2 emissions equivalent to a conventional gas-powered car.[18]

PHEVS using wind-generated electricity to recharge their batteries during off-peak hours would effectively consume gasoline at a cost of just fifty cents per gallon.[19] And since hybrids and PHEVs don't idle while sitting still—only the electric motor operates then—we could drastically reduce the total carbon emissions spewed by our vehicles.

Fuel-Cell Vehicles. FCVs use a device that converts hydrogen and oxygen directly into electrical energy, which then powers an electric motor that propels the vehicle. Since the effluent of combining hydrogen (H_2) and oxygen (O) is water (H_2O), most fuel-cell vehicles are free from harmful emissions. And since it's nontoxic, hydrogen doesn't pollute the environment.

Though abundant, hydrogen is highly combustible, difficult to confine, and expensive to harness. Since it is almost exclusively found combined with other molecules, our most effective means of isolating it—by burning fossil fuels—produces greenhouse gases. Scientists are working on ways to extract it using solar and nuclear power, but as we also have no existing infrastructure for transporting or dispensing hydrogen, it's an expensive proposition.

Despite these road bumps, nearly every major auto manufacturer has a fuel-cell program (though few vehicles are available to consumers). Cali-

> **Eco-myth:**
> Letting your vehicle idle is more efficient than turning it off and turning it back on again.
>
> **Fact:**
> Idling cars get zero miles per gallon. If you are stopping for more than ten seconds, turning off and restarting your engine is more efficient for late-model cars.

fornia uses some fuel-cell public transit buses and has twenty hydrogen fueling stations in place, with fifteen more promised soon. As engineers and scientists develop cleaner ways to harness hydrogen—from soybeans and corn, for example—hydrogen could become a potential mainstream fuel source. However, cost-effective means of working hydrogen into the mainstream transportation sector are at best decades away.

Diesel. Over the past couple of decades in the United States, passenger cars with diesel engines have fallen out of favor. Get ready for them to make a comeback. Traditional problems were high emissions of nitrogen oxide and lung-penetrating particulates, but the new generation of diesel-powered vehicles is greatly improved, and today's diesel engines are required to meet the same emissions standards as gasoline-powered vehicles.

Diesel engines are about 33 percent more fuel-efficient than gasoline engines, with diesel fuel containing 12 percent more energy than gas. In 2006, ultra-low-sulfur diesel fuel was introduced, reducing sulfur content from five hundred parts per million to a maximum content of fifteen. Biodiesel fuel, which can be made entirely from renewable biomass sources or blended with petroleum-derived diesel, can further reduce tailpipe emissions. Several car manufacturers—Mercedes, Volkswagen, and Audi among them—are reintroducing diesel cars to the U.S. market.

Flex-Fuel Vehicles. Flex-fuel vehicles, or FFVs, in production since the 1980s, use gasoline and ethanol, either exclusively or blended together. Dozens of flex-fuel models are currently available—you may even drive one and not be aware of it. Ethanol is an alcohol-based alternative fuel produced by fermenting and distilling starch crops such as corn, barley, wheat, sugar beets, or sorghum that have been converted into simple sugar. Using straight ethanol is already common in South America. Brazil, which produced almost half of the world's ethanol in 2005, supplies about 40 percent of the country's non-diesel fuel.

The most widely used fuel in FFVs, however, is E85 (85 percent ethanol, 15 percent gasoline), which is usually cheaper than gasoline, but also lower in energy—expect to get 20 to 30 percent fewer miles to the gallon. Pluses include reduced emissions of pollutants such as carbon monoxide and toxins such as benzene. On the downside, E85 increases emissions of the toxin acetaldehyde. E10—a blend of 10 percent ethanol and 90 percent gasoline—is already commonly sold at gas stations all over the United States.

AIR TRAVEL

From 1970 to 2000, U.S. air travel grew fivefold, placing aircraft as the third highest carbon emitter in the transportation sector.[20] We all enjoy the mobility that air travel provides, but in terms of your personal carbon footprint, flying adds up quickly. Precise figures vary, but a round-trip flight from New York to Los Angeles adds about 1 ton of carbon dioxide to your footprint. We're not saying you should miss your best friend's wedding, but if you were heading out of town just to get away for the weekend, it would be wise to consider a destination that's relatively closer to home.

For flights that cover short distances, consider taking the train or a bus instead. In the United Kingdom, train travel is catching up with plane travel as a preferred option, as passengers are raising environmental concerns—trains are between four and ten times cleaner than planes in terms of CO_2 emissions—and running into negative experiences concerning security screening and long delays associated with flying.[21]

CARBON OFFSETS

Most human activity—whether it's taking a private jet to São Paulo or turning up the thermostat in January—produces CO_2. Carbon offsets are buyable credits that offer people and organizations re-demption—or at least neutralization—by funding projects such as planting trees or supporting renewable energy. We don't need to tell you that the best carbon "credit" is cutting back on CO_2 emis-sions to begin with. But are organizations that sell carbon offsets providing a planetary panacea or greenwashed peace of mind?

Companies such as Terra Pass, Native Energy, and dozens of others use online calculators to help you estimate the amount of CO_2 generated by your travel and your home. You then pay to have the equivalent amount of pollution offset via carbon-reducing proj-ects. The problem is, while it's easy to compute approximate emis-

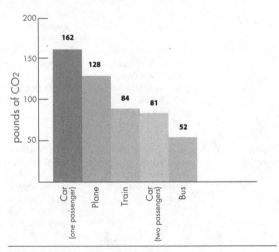

CO2 EMISSIONS AND TRAVEL
Average emissions per trip per passenger for a one-way journey
from Boston to New York (approximately 200 miles).*

pounds of CO2

Car (one passenger)	162
Plane	128
Train	84
Car (two passengers)	81
Bus	52

*All figures used are approximate. Car assumes 24 mpg.
Plane and train based on average occupancy. Bus based on 75% occupancy.

Fuels of the Future

Despite twenty-five years of effort, no alternative to petroleum has emerged as a clear front-runner to fuel vehicles. Petroleum fuels have high energy density, relatively low cost, and are supported by an extensive infrastructure. Lower-carbon fuels such as natural gas, biodiesel, and ethanol already serve niche markets and could become more widely available if tax subsidies and public acceptance continue to build. According to the Pew Center on Global Climate

FUEL TYPE	WHAT IT IS	PROMISE
Natural gas	A fossil fuel captured from wells or manufactured from organic matter such as coal.	Produces 20 to 25 percent fewer greenhouse gases and virtually no particulates compared with petroleum. Available, accessible, and transportable via existing infrastructure. Home fueling stations are a viable option.
Propane	Liquefied petroleum gas, a by-product of processing natural gas and refining crude oil.	Infrastructure for processing and transporting is existing. Much lower ozone-forming and toxic emissions.
Ethanol	An alcohol-based alternative fuel produced by fermenting and distilling cellulose or starch crops (corn, barley, wheat, etc.) that have been converted into simple sugars.	Made from renewables; produces lower emissions; can be blended with gasoline or diesel.
Methanol	An alcohol that can be manufactured from carbon-based feedstocks including natural gas, coal, and wood.	Used to make enhanced octane M85 (85 percent methanol, 15 percent gasoline), which touts lower emissions, higher performance, and lower risk of flammability than gas.

Change, by 2015 renewable liquid fuels blended with petroleum fuels could reduce transportation's CO_2 emissions by 2 percent; that number could be bumped up to 7 percent by 2030. By 2050, the Natural Resources Defense Council predicts biofuels could help could help slash transportation-related emissions by 80 percent. Below, a roundup of those most likely to succeed.

PROBLEMS	USED IN	MAINSTREAM POTENTIAL
As a fossil fuel, it is a limited resource. Though cleaner than gas and diesel, it still creates greenhouse gas and other emissions.	130,000 vehicles, including the Honda GX, transit and school buses, and refuse haulers.	Moderate to high
Flammable; made from nonrenewable resources.	Some buses; some light-duty and passenger vehicles produced before 2004.	Moderate
Shifting land use from food crops or forests to fuel crops could present problems.	Flex-fuel vehicles. Twelve percent of automotive fuel sold in the United States contains some ethanol.	High
Emits large amounts of formaldehyde.	FFVs that run on M85.	Moderate

Fuels of the Future (continued)

FUEL TYPE	WHAT IT IS	PROMISE
Biodiesel	A fuel produced from agricultural resources, typically refined soybean oil, but also canola, sunflower, and recycled cooking oils, and sometimes animal fats. B100 is pure but can be blended with conventional diesel to make B5 (5 percent biodiesel, 95 percent diesel) and B20 (20 percent biodiesel, 80 percent diesel).	Can be domestically produced and is renewable and biodegradable. B100 can reduce CO_2 emissions by up to 75 percent; reduces other pollutants, too. Requires no special equipment for fueling and most conventional diesel engines can use it without modification.
Straight vegetable oil	Food-grade waste oils, usually harvested from restaurants, but further unrefined.	Presently free; nonpolluting and nontoxic; reuses a material that would otherwise be thrown away.
Hydrogen	A gaseous element that's readily abundant in nature	Little to no back-end emissions. Advanced solar and other technology could reduce or eliminate front-end CO_2 emissions. Home fueling stations possible.

sions, actually counterbalancing them can be a tough case. Trees can take decades to reach their full carbon-sinking potential. Some die along the way. And if a forest burns down, all that captured carbon is released. Plus, is it really accurate to sell an existing wind farm as a carbon bank if it would already be doing that anyway?

Critics of these transactions say they're nothing but paper tigers that give heavy polluters carte blanche to do as they please, instead of cutting away at the real problem—reducing carbon emissions to begin with. Others warn that shady operators are simply preying on people's guilt. And because this carbon-slashing operation is voluntary, it remains largely unregulated. Because there are no federal guidelines to ensure consistency, the price per ton of carbon

PROBLEMS	USED IN	POTENTIAL
Supplies are plentiful, but not great enough to support a large-scale shift. Changes in land use—shifting from food or forests to produce fuel crops—could be problematic. Blends still rely on petroleum.	Approximately six hundred fleets nationwide.	High
Reduced engine life; not currently street legal.	Primarily DIY modified vehicles.	Low
Fuel cells and storage technology are inadequate; production is expensive and requires fossil fuels; no existing infrastructure. Highly flammable.	Honda FCX; some public buses.	Uncertain

dioxide can fall anywhere from about a dollar to twenty dollars and up. Plus, the percentage of your contribution that goes toward offsetting is left up to each company's discretion. But the airlines, political campaigns, entertainment awards shows, and individuals who are making carbon trading a $100-million-a-year (and growing) industry obviously think otherwise.

Though offsets are not a solution, they can be an intermediary step in the right direction. True, without regulation and legislation, trading schemes such as the Chicago Climate Exchange will remain ineffective against large-scale, uncaring corporate polluters. But in the meantime, something has to be done about all the extraneous CO_2 we're spewing into the atmosphere. Carbon credits are not a substitute for responsible environmental behavior, nor are they a solution to our global warming woes. However, since it's impossible to be completely impact-free, buying responsible offsets can contribute to lowering your carbon footprint.

The key to success is to ensure that your financial support results in a net improvement in the climate. To make sure a program puts your money where its mouth is, ask questions. Where will your money go? Are the projects third-party certified? What is the price per ton? How does the company calculate the carbon-capture capacity of a project and is it making exaggerated claims? Are the same offsets sold to multiple buyers, or will your funds support unique projects?

That last crucial question—the concept of "additionality"—helps ensure that companies aren't double dipping, so to speak, by hawking credits for projects that would have been implemented regardless of whether you contributed or not. Think of it this way: If the fries were already included in the price of the sandwich, you wouldn't expect to pay for them separately. You'd use that money to buy a shake.

These questions are not always easily answered. If you decide you want to buy but don't want to dedicate a portion of your year to background research, Clean Air Cool Planet thoroughly evaluates and ranks many carbon offset retailers. Swiss nonprofit The Gold Standard Foundation offers third-party certification as well. With this label sellers who work with renewable energy and energy efficiency projects are held to strict environmental criteria.

> Q: Is it more eco-responsible to invest in a new car that gets thirty miles per gallon, or to hang on to that old clunker that only gets eighteen?
>
> A: Most environmental damage from a car happens during use, not manufacture. So, generally speaking, upgrading to a significantly more fuel efficient car is the better choice.

Ultimately, whether or not to buy carbon offsets is completely up to you. You may decide the money would be better spent as a donation to an ecologically minded nonprofit or to support political change. Managing our carbon balance sheets is a little like paying off debt—at the end of the day, the goal is to be in the black. The path you take to get there will depend on your personal compass.

Q & A

Mike Millikin of Green Car Congress on the Car Industry's Crystal Ball, Driving Fifty-five, and Why the Prius Won't Save Us

You left the IT industry to create one of the Web's most popular sites about "sustainable mobility." What was the draw? The problems with transportation touch on the issues of energy, national security, and geopolitical stability. Perhaps more than in any other sector, solving these issues is dependent upon the unique intersection among companies, consumers, and exciting technology.

Can consumers significantly impact change in this arena? Or is it up to policymakers and automobile manufacturers? The most immediate and effective change we could make would come from a change in behavior with regard to the vehicles we already have—that is, ramping down fuel consumption. Well-crafted policy will help, but consumers can have a very direct impact.

Does that mean we should all be buying hybrid cars? Hybrids aren't a magic solution, and just buying Priuses won't get us out of this problem. It's really about buying time: Can we slow down greenhouse gas emissions fast enough to

Q & A (continued)

avoid climate catastrophe? We're stuck for the moment with petroleum, and technology won't come to the rescue within the critical time frame of the next ten years. Even if every new car bought today were a hybrid, we wouldn't necessarily see much change: When the price per mile of travel goes down—as it does with hybrids—people tend to drive more.

What *should* we do, then? The best thing to do is not drive. Walk or bike. Given sprawl, this doesn't work for everyone. So fundamentally, we need to get away from people driving big cars fast. Fifty percent of the fuel spent in highway driving is used to overcome air resistance. The faster you go, the more you consume. That's why the government set the speed limit at fifty-five in the 1970s. Also, don't let your car idle. In Japan, for example, people actually turn their cars off at stoplights.

What's the role of vehicle manufacturers in all this? Conceptually, they're really clear on climate change, peak oil, and the immediate need for fuel economy. But the marketing side is complicated because demand, technology, and affordability are still issues. From a corporate point of view, the market has to be tenable. I have sympathy for that. But it doesn't get them off the hook.

What technology excites you most? Advances in battery technology have been significant in the past five years. GM expects to sell a plug-in hybrid, the Saturn VUE, by 2009, and the Chevy Volt by 2010. I think they'll do very well. Efficiency improvements in combustion engines have been impressive. Sensors can tell if you're going uphill or down, for example, and adjust the engine accordingly. In the 1990s, an

Q & A (continued)

engine might have had three sensors. By 2010, it might have twenty. We're turning the engine into a thinking thing. But these are interim strategies until electric vehicles can get power exclusively from the grid, hydrogen, or other sources.

So what kind of car do you drive? I try not to drive, but I have a Prius.

From your vantage, is it doom and gloom or blue skies ahead? I'm very optimistic. People can be massively creative. But we need to create a fundamental demand shift in the market. The big challenge is to see how rapidly people can change. It's an interesting and a dangerous time.

Mike Millikin is founder and editor of the online magazine Green Car Congress.

Save the Planet in Thirty Minutes or Less

- Walk or bike to do at least one errand this week in lieu of using motorized transportation. $

- Buy carbon offsets to counter the emissions you create by traveling.

- Pledge to take at least one fewer plane trip per year. That could mean taking the train for a business trip or planning a vacation closer to home. Consider video conferencing; free software is available online. $

- Pledge to drive in a sensible, fuel-efficient manner. $

- Ask yourself "Do I need a car?" If the answer is yes, ask yourself, "Do I need a car to make *this* trip?" Don't own a car? Ask yourself if you could walk or bike instead of taking the bus or a taxi. $

reduces CO_2

improves health

$ saves money

saves time

So You Want to Do More

reduces CO$_2$

improves health

saves money

saves time

- Consider telecommuting, carpooling, taking mass transit, biking, or using a means of transport other than a car to get to work at least once a week. $

- Consider using multiple modes of transit to get around. You could ride a bike (or a skateboard or a unicycle) to the train or bus stop, for instance. Encourage transit authorities to add bike racks to buses, allow bikes on trains, add bicycle lanes to city streets, and provide more bicycle parking racks.

- Encourage others at your office to do the same. You could even start a carpooling network. $

- Considering buying a new car? Make fuel efficiency and emissions top priorities. $

- Better yet, buy used. Could you run an older diesel car on biodiesel? The National Biodiesel Board can help you learn if one of the existing eight hundred pumps is located near you. $

- Keep your car tuned up and otherwise well maintained.

- Buy a bike. Use it—a lot. $

- If you haven't already, check out the public transportation options where you live. Do they suit your needs? Could you use them more? If not, what's the problem? Better yet, what's the solution? Let your local officials know. $

- Mechanically inclined? Convert your car to run on vegetable oil. $

- Moving to a new home? Make transportation linkage a part of your selection criteria. $

- Consider going car-less. Trade ownership for services such as Zipcar. $

- Stay informed about the latest and greatest advances in automotive and bike design with TreeHugger.com's super-popular transportation news section. PlanetGreen.com can help you start a carpool.

7

Week Five

Greening Your Home:
Energy Consumption, Water, and Building

Your mission:
Save the planet one lightbulb at a time.

About 20 percent of all carbon dioxide emissions in the United States come from energy used in homes. That means that the electricity we use to light up our kitchens and power our TVs and the fuel we use for heating and cooling contribute in a pretty big way to global warming, pollution, and the depletion of our natural resources.

At 2.5 billion tons per year, the electricity sector as a whole accounts for the single largest share—38 percent—of CO_2 emissions produced in the United States. That's largely because more than 70 percent of our electrical power supply comes from burning fossil fuels—natural gas, oil, and especially coal.[1] (To find out where yours comes from, go to www.epa.gov/cleanenergy/powerprofiler).

Since 1991, regulations have helped reduce sulfur dioxide and nitrogen oxides emissions—the culprits behind acid rain—from U.S. power plants by a third. But the same success has not been achieved for greenhouse gas emissions, which have increased by

25 percent over the past two decades. Demand for electricity is also on the rise, as are operational costs, making it unlikely that power companies will voluntarily invest in technologies to reduce emissions. Coal in particular, being abundant and relatively inexpensive—it currently generates more than 50 percent of our electricity—is likely to remain a prime source of power for decades to come.[2]

Production of some lower-carbon-emitting electricity is already in place. Nuclear power, for example, which has zero carbon dioxide emissions (but other drawbacks), presently generates nearly 20 percent; hydroelectric provides another 7 percent. Other renewable energy sources, such as geothermal, wind, solar, and biomass fuel, contribute just 0.6 percent to total generation.

CONSERVATION: THE FIRST STEP

Developing renewable energy is essential to a cleaner atmosphere for not only the United States, but also the developing world. As emerging economies such as China and India become more industrialized, more populated, and richer, the types of energy they adopt will impact both their own quality of life and environments worldwide. But in addition to creating cleaner energy, we also need to reduce the amount of energy we use and, especially, that we waste. If Americans achieved a 2 percent reduction in energy use each year for the next thirty to forty years—a very feasible rate—we could get halfway to stabilizing our greenhouse gas emissions, according to Arthur Rosenfeld,

More than

two thirds

of the electric power generated in the United States is produced by burning fossil fuels, principally coal (51.2 percent), natural gas (16.6 percent), and oil (3.1 percent).

Ph.D., commissioner of the California Energy Commission. Policy changes requiring the manufacture of more efficient appliances would help us reach that goal, but things we do in our own homes—choosing a more efficient dishwasher, say, or using compact flourescent light bulbs—go a long way.[3]

We've witnessed this phenomenon in the past. Between 1975 and 2000, refrigerator efficiency increased at 5 percent per year, resulting in equivalent savings of forty one-gigawatt power plants. Energy-saving air-conditioning units have had a similar impact. Today, constructing more efficient buildings and regulating standby power used by electronics could save tremendous amounts of energy and reduce emissions.[4] But even more basic and effective would be purchasing fewer electronics to begin with. Does every kid in America need a television set in his bedroom? Do you really need a second refrigerator?

U.S. GREENHOUSE GAS EMISSIONS BY SECTOR

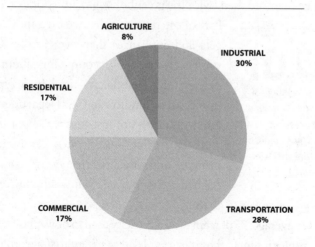

AGRICULTURE
8%

INDUSTRIAL
30%

RESIDENTIAL
17%

COMMERCIAL
17%

TRANSPORTATION
28%

Source: U.S. EPA, Energy Star Program

Conservation's potential notwithstanding, a shift to lower-carbon energies would be wise. Governmental policies to cap and trade carbon dioxide emissions and incentives for renewable energy production and use will be necessary to stabilize emissions in the short run. Ultimately, there is no one path that will lead us to a low-carbon future and a high quality of living. A combination of many technologies—coal with carbon capture and sequestration; increased efficiency in the generation, transmission, and end use of electricity; increased use of lower-carbon fossil fuels such as natural gas; renewable, nuclear, and decentralized power generation—will likely be used to pave the way to a greener future. Political legislation and personal and industrial behavioral change remain a requisite part of the mix.

PHANTOM POWER

For most of us, our homes are the biggest contributors to our personal carbon footprints. Energy used in the average household can equal more than twice the greenhouse gas emissions of a car.

70:

Percentage of Americans who believe it's possible to reduce the effects of global warming.

Part of the reason is that we own more gadgets and appliances than ever before. Consumer electronics account for about 20 percent of all electricity used in a typical home, according to the U.S. Department of Energy. But it's not just that we've got more stuff. Our cable boxes, video game cubes, printers, and other electronics are continuously drawing power, even when they're not in use.

On average, 75 percent of all electricity used to power household electronics is sent to these devices *while they are turned off,* according to the U.S. Department of Energy. We actually spend more money to power our DVD players and audio equipment, for example, while they're in standby mode than while they're actually entertaining us.[5] Severed from their better halves, yet left plugged into walls, chargers for things such as cell phones, iPods, and cameras continue to suck energy even while not attached to their mates. With the national average at five chargers per person, this adds up. Rechargeable battery docks for things such as drills and handheld vacuums can draw from five to twenty times more energy than is stored in their batteries. This squandered electricity, known as "phantom" or "vampire" power, can be conserved by unplugging equipment or using a power strip that can be turned off, making a huge difference.

CHOOSING APPLIANCES

Using more-efficient gadgets and appliances to start with, of course, also results in less energy splurging. While some of these may cost more up front, many pay off in the long (or short) run. Though much more could be done, some of the most effective approaches to promoting efficiency have involved federal standards and labeling programs, and utility companies that have created incentives for efficient end use.

The Energy Star program—a labeling initiative enacted by the U.S. EPA and the Department of Energy—designates appliances that beat federal conservation and energy-use standards and cost less to run. Energy Star–rated consumer electronics in standby mode use as much as 50 percent less energy than conventional equipment. (We still recommend unplugging gadgets or using power strips whenever possible.) When it comes time to replace electronics and appliances, choose Energy Star–rated items, which indicate greater efficiency. If your fridge was manufactured before 1993, for example, an Energy Star–qualified model could consume half the energy, saving you money and about 100 pounds of CO_2 a year. (Note that the Energy Star and EnergyGuide labels are not the same thing. The latter is a yellow sticker that is not an indication of high energy-efficiency, but rather shows an appliance's estimated energy use compared to similar models and estimated annual operating costs.)

LIGHTING

Household lighting accounts for about 12 percent of home energy consumption and produces more than a ton of carbon dioxide each year.[6] You probably already know that changing out incandescent lightbulbs for compact fluorescent lights, or CFLs, is the eco-conscious way to go. Incandescent bulbs turn only about 10 percent of the electricity they use into light; the rest is wasted as heat. (Halogens produce even more heat.) CFLs use approximately one third the energy of traditional bulbs and can last up to ten times longer. They sometimes cost a little more up front, but Energy Star predicts the payback can be as big as thirty dollars per bulb over its lifetime.

It used to be that CFLs were flicker-y and unappealing, but they've come a long way. While some still take a few moments to work up to full brightness, most come on without a hitch and provide high-quality light. True, fluorescent lights might not be perfect for every situation, but replacing your high-use bulbs with energy-efficient models is an effective defense against global warming simply because it is so easy to do and has such a big payoff. If every household replaced just one bulb, we'd save enough energy to light more than 2.5 million homes for a year and prevent greenhouse gas emissions equivalent to taking nearly eight hundred thousand cars off the road.[7] CFLs make so much sense that the governments of Australia and Ontario have passed legislation to ban incandescent bulbs.

Another lighting alternative is LEDs, or light-emitting diodes. Though they are presently significantly more expensive, LED bulbs can reduce energy consumption by up to 90 percent and will last about a hundred thousand hours. LEDs are extremely bright, light up very quickly, and are already widespread. Common applications include car brake lights, desk lamps, and holiday string lights. Many products with LED lights built in are reasonably priced. Affordable LED bulbs that fit a wide variety of typical fixtures can't be far behind.

Break Cover: Safe CFL Clean-up

Unlike incandescent bulbs, compact fluorescent lights, or CFLs, use small amounts of mercury to operate. Some concern has arisen over this, but according to life-cycle assessments, far more mercury is released from the generation of power used for incandescent bulbs than in the manufacture and use of CFLs over the course of their working life. Still, mercury is an extremely toxic pollutant, and consumer responsibility is essential. Compact fluorescent lights are recyclable into their component parts—aluminum, glass, calcium phosphate, and mercury—which can be recovered and reused in new bulbs. (Fluorescent tubes, like the overheads in an office, are still tricky to reuse, however.) Proper disposal of CFLs is extremely important. Ikea already has a strong take-back program; expect recycling centers to pop up soon at big-box stores such as Wal-Mart and Home Depot.

If you're worried about breaking a CFL in the house, know that exposure from a shattered one is not likely to pose much risk. Since CFLs contain relatively small amounts of elemental mercury—about four to five milligrams, typically—the compound is likely to vaporize and disperse from a room within a few hours. Since mercury can enter the body via the skin, mouth, or

---

Break Cover: Safe CFL Clean-up (continued)

lungs, however, it would be wise to minimize any potential exposure.

To do so, the EPA recommends immediately opening windows to reduce volatized mercury concentrations, avoiding touching or handling the mercury, cleaning up broken glass carefully and im-mediately—but not with bare hands or a vacuum cleaner—and wip-ing the affected area with a paper towel to remove all excess glass fragments and mercury. Finally, place the paper towel and bulb remnants in a sealed plastic bag and dispose of it at a hazardous waste collection site.[8]

AIR CONDITIONING

An astounding one sixth of all electricity produced in the United States is used to air condition. Houses with central air can reduce strain on the grid by keeping vents and ducts clean. Programmable thermostats—which work for heating, too—help regulate tem-perature while you're away but ensure you'll be in the comfort zone upon arriving home. For in-window units, buying the proper size unit for the space where you'll use it should be a top priority. To save even more energy, turn the A/C's thermostat up a couple of degrees. Better yet, avoid using air-conditioning alto-gether whenever possible. Ceiling fans work in conjunction with air-conditioning and heat to keep spaces cool or warm by circulat-ing air; but since they don't change the temperature, they should be turned off when you exit a room. A wiser option for people who live in dry climates is an evaporative cooler (sometimes called a desert cooler), which distributes cool moisture into the air using less energy. Better still, designing structures to have proper cross-ventilation from the start can help reduce or eliminate the need for air-conditioning.

---

thermostat down two degrees in winter and up two degrees in summer: 2,000

HEATING YOUR HOME

In the United States, buildings account for more than 40 percent of all energy used and pollution generated.[9] Part of the problem is buildings that leak heat and air-conditioned air as if that's their job. The majority of households in America are heated with oil, natural gas, propane, or electricity. While some of us may have the means and ability to change these circumstances, many of us are stuck with what we've got. Whether you own or rent or you live in a hot, cold, or variable climate, if you use heat or air-conditioning, weatherizing your home to reduce draft helps keep climate-controlled air where it belongs—inside.

The gaps around windows and doors in an average American house are equivalent to having a three-by-three-foot hole in the wall, according to the Natural Resources Defense Council. Insulating is the best step you can take, followed by using caulking and weather stripping to seal off air leaks; heavy drapes and other insulating window treatments will also provide more insulation. (Don't forget that the better you seal up your home, the more you'll need good ventilation.)

The wind in North Dakota could produce **a third** of America's electricity.

Supplemental heating sources can be cozy, but aren't always efficient. Fireplaces can allow a drastic amount of the heat in your home to escape through the chimney. If you can't resist using yours, dropping the thermostat will stop your heating system from trying to replace warm air being lost. Installing glass doors on the fireplace, which can be closed when the fire's dying or out, will also prevent warm indoor air from escaping, as will closing the chimney damper when the fireplace is not being used. An outdoor air intake will also cut down on heat loss, and

ENERGY USE AT HOME
The average annual energy bill for a typical single family home is about $1,900.

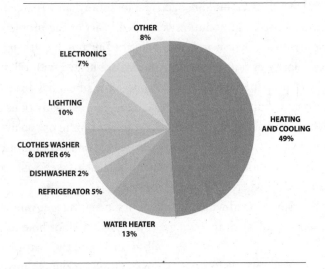

Source: U.S. EPA, Energy Star Program

a high-efficiency fireplace insert offers stricter air control. Better options include wood-burning or pellet stoves—these use wood, switchgrass, corn, or other biological materials—which are highly efficient and don't allow air exchange between inside and out.

New furnaces and boilers are required to run at much higher efficiency than previous ones. (Before you get a new one, though, make sure your old one can't be improved by reasonable modifications or repairs.) While natural gas boilers and furnaces, as well as other natural gas appliances, in general produce a lot less CO_2, furnaces that use oil can run on biodiesel. Several companies now supply bio-blend heating oil, sometimes called Bioheat, to residential customers, who may be eligible for tax incentives in some states.

ceilings insulated? Add insulation if they're not: 2,000 • Install a low-flow

Passive Solar Energy

Passive solar energy is an excellent example of how smart design can provide a simple solution to big-picture problems such as energy efficiency. Passive solar energy techniques such as solar thermal mass systems don't necessarily require special technology, just good design. Simply put, solar thermal systems absorb radiant heat from the sun then release it when the sun is not shining. Walls or floors of concrete, brick, or tile can be used to store heat and regulate indoor air temperatures. In winter, thermal mass absorbs heat from direct sunlight, releasing it when the sun goes down. In summer, keeping thermal mass shaded allows it to draw warmth from the air, thereby cooling a house.

Geothermal Exchange Systems

Ground-source heat pumps, also called geothermal exchange systems, take advantage of constant temperatures of about 50 degrees Fahrenheit just a few feet underground. In summer, an electric pump moves liquid through pipes to remove heat from a building; in winter, the pipes deliver warm air. Particularly suited for regions with temperature extremes, geothermal pumps are up to 70 percent more efficient than traditional HVAC systems and much less polluting. Though they have higher up-front costs and can be difficult to install in existing buildings—since wells must be dug for the piping—geothermal heat pumps can save homeowners hundreds of dollars each year and are excellent op-

> **Eco-myth:**
> It is more efficient to leave the lights on if you're leaving a room for only a short period of time.
>
> **Fact:** For standard incandescent lights, the rule of thumb is to turn them off, even if you're leaving the room for only a few seconds. With CFLs, turn them off if you'll be away for more than three minutes.

tions for new construction. Today, about five hundred thousand geothermal heat pumps have been installed in the United States— George W. Bush reportedly even uses one at his Texas ranch.[10]

Combined Heat and Power

Combined heat and power (CHP) units have long been used in institutional settings, but practical constraints have kept them out of the residential market until now. Already popular in Japan and Europe, this technology, which simultaneously creates heat and electricity, is suited primarily for colder climes. When there's a call for heat or hot water, a natural gas– or propane-powered engine runs while a co-generator captures excess heat, which it then turns into electric power. When heating requirements peak, a furnace or boiler augments the engine's output, producing yet more power. If electricity production overshoots consumption, surplus is sent to the grid. By converting 83 to 93 percent of fuel into energy, compared with the 30 to 40 percent efficiency achieved by conventional central power plants, micro-CHPs result in significant emissions savings as well: An average home can cut its roughly 20,000 pounds of carbon dioxide emissions down to 5,000, according to Climate Energy, the first company to install such systems in the United States.[11] Additional bonus: The micro-CHP sits stealthily in the basement, an advantage over outdoor eyesores.

CONSERVING WATER

Water is arguably our planet's most valuable resource, and we're using it up fast. Demand for water has tripled over the past fifty years, aquifers are shrinking, rivers are running thinner, and wells are going dry.[12] Some experts warn that in the near future, wars will be fought over not oil but water. Some argue that they already

WATER USE INSIDE THE HOME

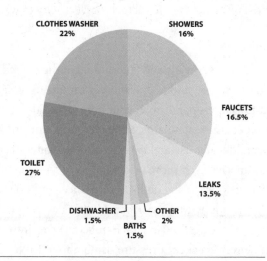

Note: Percentages are approximate due to rounding.
Source: U.S. EPA, Office of Water

are, in places such as Darfur. On the flip side, we have trouble dealing with excess water, too. Storm runoff is polluting our rivers, streams, and oceans, and heavy rains can cause sewage overflow. How can we rethink our approach and reengineer our communities in order to strike a balance? To start, water conservation is essential to shaping a sustainable way of life.

The average American uses more than 80 gallons of water per day, while the rest of the world averages just 2.5 gallons per person.[13] You can cut back significantly, without sacrificing your quality of life, simply by using efficient fixtures and appliances, which will also save you money on your water and water-heating bills. At about 13 percent of overall energy use, heating up water is typically the third largest energy expense in a household, and comprises nearly 4 percent of our country's total energy consumption.[14]

(Sources: StopGlobalWarming.org, Rocky Mountain Institute)

Using less hot water also helps reduce global warming emissions, since most residences use fossil fuels or electricity to warm it.

Showers

As of 1992, showerheads and faucets sold in the United States are required to use 2.5 gallons of water per minute or less; toilets are allowed 1.6 gallons per flush. If your home has older fixtures, they may not meet these standards. Showering accounts for 21 percent of household water use and two thirds of all water heating costs, but you can save even more water and money—even if you have newer fixtures—by using high-performance, low-flow, and aerating showerheads, without sacrificing beloved water pressure.

Aerating showerheads, which mix airflow into the water stream, deliver excellent pressure using about 1.0 to 2.2 gallons per minute—some, even fewer. Others feature "pause" buttons that hold water back while you soap up or shave. In the bathroom, a faucet that delivers 0.5 to 1.0 gallon per minute should be sufficient. In the kitchen, stick to 2.0 or 2.5 gallons per minute, which makes filling pots easier.[15]

Water Heaters

Turning down the thermostat on your water heater from 140 to 120 degrees Fahrenheit (still plenty hot) also saves energy. Insulating your hot water heater is another cheap and easy way to make it more efficient. Most hardware stores carry water-heater blankets, which cost about ten to twenty dollars and can reduce heat loss 25 to 45 percent.

Water heaters more than ten years old are likely running at less than 50 percent efficiency. If yours is, replace it with an energy-efficient model or, better, switch to a tankless heater, which warms up water only as you need it rather than holding hot water at the ready all the time.

Solar water heating systems, which can be used in virtually any climate—having a sunny, south-facing location for placement helps—can typically meet 50 to 80 percent of most households' hot water needs. Though they cost slightly more to install than traditional systems, your fuel source—the sun—is free. One solar water heater can save more than 50 tons of carbon dioxide emissions during a twenty-year period and helps reduce reliance on fossil fuels.[16]

Doing the Dishes

Ever wondered which is more efficient, washing by hand or using the dishwasher? Good news: Using the dishwasher is actually better for the planet. The average energy-efficient dishwasher uses just four gallons per cycle, saving up to 5,000 gallons of water per year compared to hand washing, as well as $40 in energy costs and 230 hours of your time, according to the EPA. You can further increase efficiency by scraping instead of rinsing plates before loading them, running only full loads, and choosing the energy-saver or light wash and air-dry options. If *you* are the dishwasher, fill the sink instead of letting the water run, and rinse as efficiently as possible.

Gray Water

Household wastewater from sinks, showers, dishwashers, clothes washers, and other sources is called "gray water" and can be collected and reused to irrigate ornamental landscaping plants (it shouldn't be used on plants intended for consumption) and sometimes in toilets. (Sewage water, or "black water," must be treated and sanitized.) Regulations regarding the use of gray water widely vary by community—some require it to be treated, others don't, and some don't allow it. Recycling water in this way can be achieved by installing simple or complicated plumbing systems, but it can also simply mean emptying used cooking or fish tank water into the garden instead of down the drain.

Toilets

The amount of water we flush down the toilet is astounding. At twenty-two gallons per day, it accounts for 26 percent of all household water use, our largest source. Dual-flush (which have separate settings for liquid and solid waste) and low-flow (which use a minimal amount of water) toilets surpass government regulations. You can also replace just the handles and tank. Even more basic, placing a brick in your toilet tanks replaces the same volume of water, meaning less is used each time you flush.

Laundry

Washing machines are the second biggest water-hogging appliances, using 22 percent of household H_2O. Standard washes can use as much as thirty-two gallons per load, though front loaders can reduce usage by 40 percent and use up to 65 percent less electricity. Using the Laundromat gives you an eco-advantage; industrial washing machines—dryers, too—tend to be more energy efficient.

Up your eco-quotient by washing clothes in cold water and running only full loads. Wearing clothing more than once before tossing it in the laundry bin also reduces the number of loads you'll do overall. We're not suggesting that you wear dirty socks or stained T-shirts; just consider if those jeans you had on for only two hours could sustain another outing before you start the spin cycle. Also, skip the dryer and line-dry instead; you'll improve your carbon footprint by hundreds of pounds each year.

LOW-CARBON TECHNOLOGIES

In 1850, about 90 percent of our energy came from renewable energy sources. Today, that figure has dipped to less than 10 percent, while more than two thirds comes from burning fossil fuels; nuclear energy rounds out the remainder.[17] Though technology exists that could stabilize and even reduce greenhouse gas levels within a few decades, current infrastructure, political policy, and public resistance have been roadblocks to change. In order to understand the importance of energy consumption, it's critical to know where our energy comes from—and where clean technology could take us.

Carbon Capture and Sequestration

We may be heading toward peak oil, but the end of coal is nowhere in sight. Yet to deal with climate change, we must address emissions from coal-fired power plants—and fast. To do so, experts are racing to figure out how to build facilities that can (a) get more bang for the buck and (b) capture and sequester carbon dioxide for the long haul.

Coal's upside is that it's a fairly flexible fuel. Though it's traditionally burned, it can also be converted into a synthetic gas via a process called gasification. In this form, it becomes fairly easy to remove pollutants such as mercury, sulfur, and particulates. The gas can then be reconstituted into various different liquid fuels, which can be used for home heating or in vehicles. This process, called coal-to-liquids, is not widespread and is still expensive.

A second process, known as an integrated gasification combined-cycle system, substitutes the synthetic gas for natural gas and may hold more promise. An IGCC power plant creates much cleaner electricity than standard coal facilities because it uses less coal to create an equivalent amount of power. It also makes it easy to capture the still-problematic amounts of CO_2 it does spit out.

The big problem is, no one knows where to put the collected carbon. The leading scheme is to pump it into underground reservoirs, but whether that will actually work is anybody's guess. Still, though greenhouse gases aren't regulated yet, most power providers assume that legislation capping carbon emissions is on the horizon. And for that reason, a handful of IGCC plants can be seen there, too.

Solar Power

The amount of solar energy that reaches the United States each year is 3,900 times our power needs.[18] Solar thermal electric power plants generate electricity by converting the sun's radiation into heat that then drives turbines or engines. In the United States, these plants generally use parabolic trough, dish, or power towers to collect the sun's rays and heat up a conductive material, which is then used to power an engine or turbine to make electricity. Though solar electricity generation has been expensive to produce, and power cannot be generated when the sun is not shining, it is a promising technology.

Wind

Wind is plentiful, clean, and free—which is why it's the fastest-growing energy source in the world. By turning the kinetic energy of air currents into electrical power, we could have the potential to supply 20 percent of American electricity demand with capital costs that are far lower than those of solar and comparable to those of many fossil fuel plants.[19] Whether wind turbines pose a major threat to bird populations is an important question. There is no doubt that wind turbines do kill some birds. However, the number is a tiny fraction of the hundreds of millions killed by crashing into windows

Eco-myth: Wind turbines are extremely dangerous to birds.

Fact: Run-ins with domestic cats and windows kill hundreds of millions of birds each year; wind turbines affect only a small fraction of that number.

and by cats each year. Birds are also affected by air pollution; wind power would ensure a healthier atmosphere for them. With proper siting to avoid migration routes and nesting grounds, wind turbines pose little threat to most bird populations. Where birds would suffer greatly, wind farms should not be built.

Biomass, Biogas, and Biofuel

Biomass fuel is derived from plant and animal matter. Because the carbon in plant material is extracted from the atmosphere during photosynthesis, using biofuel essentially produces no net carbon dioxide emissions. Precise emissions vary depending on how crops are grown, since fossil fuels are used for cultivation, transport, and so on.

"Biomass" typically refers to the organic matter used to make biogas, or biofuel, produced by the decomposition of organic matter, which produces methane. Common biomass sources include crops such as switchgrass and corn, food and slaughterhouse waste, manure, sewage sludge, and municipal solid waste. Biomass power plants use biofuel to generate electricity; by-products of the process are steam and hot water. Concerns regarding production of biomass include clearing land for agriculture and monoculture crop issues.

Landfill Gas

As landfill material decomposes, it produces a combination of carbon dioxide and methane called landfill gas, or LFG, a type of biogas. Accounting for about 25 percent of all anthropogenic methane emissions in the United States, landfills are our greatest source of methane. Methane must be controlled as it is released in order to reduce fire and explosion hazards.[20] Some simply burn LFG away, but more than four hundred U.S. landfills are extracting, capturing, and refining it to produce electricity, to replace fossil fuels in industrial and manufacturing operations, or to be upgraded to pipeline-quality gas. The EPA estimates that more than six hundred additional landfills could support similar projects cost-effectively. For every 1.0 million tons of municipal solid waste, these facilities generate enough energy to power 725,000 homes, heat 1,200,000 homes, and keep annual emissions from the equivalent of thirteen million vehicles out of the air.

Geothermal Power

The amount of heat within about six feet of the Earth's surface contains fifty thousand times more energy than all the oil and natural gas resources in the world. Geothermal energy taps into that potential, using naturally occurring hot water or rocks to create steam, turn turbines, and generate clean electricity. Though geothermal heat occurs everywhere, ideal conditions are found in less than 10 percent of land worldwide; areas with active or geologically young volcanoes—such as many western states, including California, Oregon, Nevada, and Alaska—are prime locations.[21]

Like fossil fuels but unlike most other renewable energy sources, geothermal power can supply continuous base-load power. Though it is the world's largest producer of geothermal energy, the United States currently capitalizes on only a fraction of its potential capacity. About 2,800 megawatts—the energetic equivalent of burning about sixty million barrels of oil—are produced here today; that number is expected to leap to between 8,000 and 15,000 megawatts over the next decade as prices become more competitive with fossil fuel–derived electricity.[22]

Hydropower

Hydroelectric power was historically considered environmentally benign and carbon-friendly (though decaying organic matter trapped by dams does emit some CO_2), but today flooding and damming large areas is known to throw aquatic ecosystems seriously out of balance. Added to the fact that we've tapped most potential sources, traditional hydropower, which accounts for about 7 percent of our national power supply, is unlikely to grow further.

89: Percentage of TreeHugger staffers who believe nuclear power is not a good solution to our energy needs.

New ideas in hydropower, however, are on the horizon. Using the constant motion and power of water—that is, the "hydrokinetic" energy of running rivers, changing tides, and rolling waves—we can move hydroelectric turbines. Though ecological side effects haven't been fully studied and cost-effective approaches haven't emerged, harnessing the power of water to generate electricity is promising. In places such as New York's East River and

60: Percentage of TreeHugger readers polled who think it is.

Northern Ireland's Strangford Lough, turbines are already testing these innovative ideas, with varying degrees of success.

Nuclear Energy

At nearly 20 percent of our nation's energy production, nuclear power is our largest source of carbon-free electricity. Advocates of increasing nuclear energy production suggest that it's the best—perhaps the only—option for meeting rising power demand while reducing our carbon count at a price that is competitive with coal and natural gas. Pricing would become even more competitive if the government were to implement a carbon tax, which would drive up the cost of energy generated by fossil fuels.

But going nuclear is a tough sell for critics, who point out that nuclear power plants are difficult to site and costly to build, and that radioactive and weapons-grade plutonium—by-products of nuclear power production—are challenging to control.

Though decades of study show that geologic disposal of nuclear waste—that is, burying it below ground—has proven effective, it's impossible to know how well this will work over the long haul. And no one has yet come up with a system for permanently disposing of spent fuel and other radioactive waste, which takes thousands of years to break down fully. Furthermore, one study found that in order for nuclear power to make a dent in global warming, we'd need to build three thousand reactors—that's one a week—over the next sixty years.[23]

A severe nuclear accident—which some fear is an accident waiting to happen—could do catastrophic harm to humans and the environment, potentially killing and injuring thousands of people, contaminating enormous areas of land, and costing billions of dollars. Deliriously frightening disasters at Chernobyl and Three Mile Island have left a black mark on public opinion regarding nuclear power, though it should be noted that in the case of the latter, the containment structure did its job—no one died, and no radioactive substances were released into the environment. Those in favor note that nuclear accidents have been few and far between, that a new generation of facilities would be safer than those before, and that air pollution from traditional power plants kills far more people than nuclear accidents ever have. Ultimately, whether or not to support increased nuclear power is a difficult call—one that splits the environmental community.

BUILDING GREEN COMMUNITIES

During the past three decades, the average American home has more than doubled in size from less than 1,000 square feet to about 2,400 square feet, while family size has shrunk.[24] At the same time, we are living farther from where we work, shop, and entertain ourselves. Building homes using greener materials is important, but to truly build green, we must look at how a residence functions with its community. Is it part of multi-housing unit with shared utilities, or is it a stand-alone? Is it located within walking or biking distance to the grocery store, workplaces, and parks? Does it decrease its occupants' reliance on using automobiles and are public buses or light-rail accessible?

Even a house built from the greenest materials doesn't accomplish its goal if it is enormous (bigger houses necessarily use more energy) and if residents must make long commutes to get to work or do daily errands. To that end, we must build more houses that are clustered, share utilities, share common resources, and allow for more open space. Continuing to build detached houses in isolated sub-developments will compound environmental problems. In Europe, where for centuries houses have been built around town centers where residents shop, work, and socialize, communities generally have a smaller environmental impact.

Co-Housing

Concepts such as co-housing, first developed in Europe and now gaining exposure in the United States, explore ideas of community-based development. Residents of co-housing communities keep dwellings with private spaces—kitchens, living and dining rooms, bedrooms—but share common facilities that most people don't need to use every day—entertaining and lounge areas, meeting rooms, guest chambers, recreation facilities, libraries, and workshops, for example. Residences are often built around open spaces, and amenities such as day care or exercise classes may be offered on-site.

Green Architecture

Development and planning issues are paramount and emerging building technologies and materials are making it easier

Prefabricated Housing: Hip, Modern, Green

Prefabricated Housing: Hip, Modern, Green

Partially built in factories then shipped in parts to be constructed on-site, prefab houses cause minimal site disturbance, reduce construction energy and waste, and result in shorter construction times. Because several modernist architects have taken an interest in it, much prefab has been elevated above its drab reputation of yesteryear. Because designs are standardized, layouts are generally well resolved and efficient, offering open, flexible floor plans that result in a smaller ecological footprint. Plus, prefabricated dwellings can offer architect-grade design for a lower price.

for architects, designers, and contractors to create neighborhoods and buildings with better energy-efficiency and indoor environmental quality. (Just as your furnishings and décor can off-gas harmful chemicals, so can building materials.) But often the greenest building concepts are also the simplest: Building with the environment in mind starts with designing smaller houses that are properly sited. (Generally speaking, rehabbing existing buildings is also more eco-friendly than building new.) Such structures use fewer resources, cause less disruption to local ecosystems, and require less energy and water to operate. Intelligently designed smaller homes and apartments with flexible spaces—for instance, a kitchen-dining area separated by a moveable partition that can allow residents to modify the space for entertaining large groups—eliminate the need for extra rooms that must be heated, cooled, lit, and maintained even if rarely used.

LEED for Homes

The U.S. Green Building Council is a non-profit organization that works to promote environmentally responsible construction. Through its Leadership in Energy and Environmental Design program, the USGBC offers voluntary ratings that certify buildings as green in the commercial, institutional, and residential sectors. LEED for Homes, which launched in 2007, verifies that residential buildings meet various criteria such as siting for solar orientation, indoor air quality, transportation linkages, and homeowner awareness. Of course, a house doesn't have to be LEED-certified to be eco-friendly. Indeed, many houses go above and beyond LEED standards without ever applying for certification. But the USGBC is a good resource for green building information—whether you're building new, adding on, or renovating—and is helping educate the public and build awareness.

Construction Techniques and Materials

There's no one right choice when it comes to building green; many different construction options exist, from standard to alternative practices. The best choices for any particular project—from new construction to small additions—will depend on where you live and what your individual objectives are. Some conventional building techniques and architectural styles can be adjusted for greener results, but choosing smarter materials and more efficient mechanical systems are always wise environmental investments that can often help you save money.

Framing. A range of framing materials and techniques are available for building, from factory pre-made structural insulated panels to high-tech wood-framing systems to straw bale construction. Which to choose depends on goals, budget, and region, but basic eco-friendly framing features should include a combination of a reduction in the amount of materials used, the creation of a structure that supports excellent insulating properties, and the use of materials manufactured in an ecologically sound way.

Insulation. In cooler climates, insulation is the number one way to make a dwelling more efficient. Any insulation is better than none, but ecologically superior types use materials that leave a lighter footprint and will ensure better indoor air quality. A few notable types include Biobased Insulation, a polyurethane spray foam made from soybean oil; Bonded Logic's UltraTouch, fabricated from post-industrial denim; and Icynene, a formaldehyde-free soft-foam insulation.

Roofing. Typically, the less often a roof is replaced, the more eco-friendly it is, so the goal should be to build one that lasts as long as possible. Petroleum-derived asphalt shingles are inexpensive, but can pollute rainwater runoff and aren't as durable as materials such as clay, concrete, or aluminum. Also, a light-colored or white roof will increase reflectivity, reducing the need for interior cooling in hotter months. "Green roofs" use drought-resistant plantings to increase insulation and reduce cooling needs and reduce storm-water runoff.

Overhangs. Overhangs protect interiors from blazing summer sunshine but also allow light in during winter months, when the sun is lower in the sky. Depending on the climate, they can help reduce energy bills or, in some areas, eliminate the need for heating and cooling altogether. If you can't add overhangs, awnings are another option.

Siding. While aluminum and vinyl siding are durable, are sometimes made from recycled materials, and eliminate the need for painting, they can only be downcycled or landfilled at the end of their useful life. Better choices may be sustainably harvested wood or fiber-cement shingles or clapboards. Depending on location and budget, stucco and brick may also be options.

Windows. Inefficient windows can be a source of massive energy loss. New high-performance windows can include reflective coatings and trapped gases that improve thermal properties by deflecting or increasing solar gain at different times of year.

Windows that provide plenty of natural daylight help reduce electricity usage. When carefully placed, they also help achieve or avoid passive solar gain. For example, a home in New England would benefit from south-facing windows that collect winter sun; a house in Arizona would keep these to a minimum to reduce cooling requirements. If new windows go against your budget or your architectural aesthetic, insulating window treatments and storm windows are an excellent option.

Decking. FSC-certified or salvaged hardwoods are good choices, though hardly maintenance-free. Made from composite wood and plastic, durable decking such as that made by Trex features recycled content but can only be downcycled or landfilled at the end of its useful life.

Shrubbery and trees. Shrubs—particularly evergreens—provide extra wintertime insulation and protection from wind, as do trees, which also create shade in summer and reduce the need for air-conditioning.

Permeable surfaces. Allowing hard surfaces, such as driveways and patios, to also be pervious means that water can seep into the subsoil, preventing toxins in runoff from entering the municipal wastewater stream. Many products, such as ECO Pavers concrete blocks by EP Henry, interconnect like puzzle pieces to create a durable surface that is also permeable.

Rainwater runoff. Proper water drainage is essential to keeping your house in good shape and ensuring fewer resource-intensive repairs. Rain barrels are one option but can be aesthetically unappealing or create insect and mold problems. Rerouting water away from a house, via a French drain or other method, and into a "rain garden" containing water-loving plantings earns you extra eco-credit for reuse—and style points, too.

Solar panels. Residential solar panel arrays—also called photovoltaic, or PV, systems—can be either roof- or ground-mounted or sometimes directly integrated into a roof. PV systems can stand alone, requiring batteries to store energy, or be tied into the grid so that when a house generates more electricity than it uses, its meter spins backward and sends the extra power to the grid. This method, known as net-metering, helps relieve stress on central power plants and earns homeowners credit on their utility bills. That solar arrays are too expensive and cannot provide enough power for a typical American home is a typical misconception. The truth is that employing conservation techniques—using CFLs throughout a residence, for example—and building smaller homes that necessarily require less electricity makes solar an obvious and intelligent option.

Home wind turbine. Home wind turbines come in all manner of sizes and vary widely in cost and output. Many require at least one-half acre of land for installation due to zoning laws, though some rooftop versions are also available. Like photovoltaic systems, home wind turbines can and should be set up for net-metering where available. Without battery backup, a wind system will not power a house during an electricity outage or when the wind is not blowing, another good reason to be tied into the grid.

Landscaping and Lawn Care

Nationwide, the water we use to drench our lawns and gardens accounts for one third of all residential water use, totaling more than *7 billion* gallons per day. As much as half of that is wasted due to evaporation, runoff, or overwatering.[25] Green landscaping means

Landscaping and Lawn Care (continued)

efficient watering, planting, soil building, and water runoff management, which will also benefit your garden. Of course, the greenest lawn is no lawn at all. Could you use the space to plant a fruit and vegetable garden instead? Our yards are corridors to our natural habitats; treating them as such can minimize their environmental impact on the world within and beyond our fences.

Choose nursery-propagated native plants. Because they are already adapted to the environment where you live, they thrive with minimal care, requiring less water, fertilizer, and pesticides. Xeriscaping, a landscaping method that uses native and drought-resistant plants, is an excellent approach, especially in dry climates such as Arizona and California. Also avoid invasives. Their seeds can be carried by birds into wild habitats. If grass is your thing, however, follow these tips:

- Minimize your yard's footprint by choosing a type suitable for the area. Use organic fertilizers and consider groundcover instead of grass in areas that don't get much foot traffic. This should require less water and care and won't need to be mowed.

- If you must water your lawn, you likely don't need to do so everyday. The EPA recommends stepping on a patch of grass to test its water content; if it springs back, step away from the hose. A well-maintained irrigation system can help ensure that water is distributed evenly and appropriately, and watering during the coolest parts of the day can help minimize evaporation.

- When it's time to trim, push mowers help keep the carbon count down. Next best choice: an electric mower. (Realize, however, that these draw energy from a polluting grid.) Cutting to the proper height also saves on water and reduces the need for fertilizer. Leave grass clippings where they fall, which keeps yard waste out of landfills. Mow over leaves, too, to achieve the same benefits.

- Chemical fertilizers have a salt base, which causes a pH imbalance in most soils and destroys beneficial microorganisms. They're also full of harmful nitrates that seep into the groundwater and beyond. Organic mulch and compost, on the other hand, provide nutrients, thereby reducing the need for fertilizer to begin with and produc-

Landscaping and Lawn Care (continued)

ing healthy grass and plants that are naturally more pest- and disease-resistant. They are also healthier for kids and pets. Mulch will help soil retain moisture, among other benefits.

- When it comes to pest control, always start with the least-disruptive and least-polluting options before resorting to stronger methods. Herbal pest repellants—such as garlic and hot-pepper sprays, a few drops of soap, or other botanical pesticides—get the job done without polluting the environment.

Q & A

Emily Mitchell, of the U.S. Green Building Council, on the Importance of Energy Audits and the Cost of Going Green

What is LEED for Homes? LEED stands for Leadership in Energy and Environmental Design. It's a voluntary program of the U.S. Green Building Council to advance market transformation in the building industry. It began with commercial buildings, but we launched the residential program in fall 2007 because this sector comprises such a large portion of environmental impact. We're also looking at community-scale projects.

What are the basic principles? It's about looking at home building from a comprehensive level—the various credit categories include energy, water, and resource efficiency; materials; and indoor environmental quality. Siting and access to amenities and transportation are important, too. We also reward homeowner education, because a big portion of the way that a home performs is based on how its occupants operate and maintain it.

Q & A (continued)

It seems like a lot about green building—proper solar orientation, using overhangs, and energy-efficiency, for example—is less about fancy technology and more about a return to traditional common sense. A lot of the LEED strategies are just smart practices that have been around in some form since early home building. Plenty of builders are using a lot of the elements that we reward, without even realizing that they are engaged in a form of green building.

So, it's not just about installing solar panels or having a home with a wind turbine? Exactly. Certainly advancements in technology—especially in mechanical systems and renewable energy—have allowed us to push forward. But a big emphasis of LEED for Homes is taking a comprehensive, integrated approach. It's not about a particular advanced technology but the way in which several different types of technologies and, more important, intelligent strategies are packaged together. There is a multitude of ways to make that happen.

That said, what *are* some of the bleeding-edge technologies you're seeing? Is everybody going to have geothermal heating and cooling by 2015? Some of the deepest green homes are modular and factory-built models. Since they're partially pre-built, they have very high efficiencies in framing.

Is the LEED designation available for renovations? LEED for Homes is for new residential construction as well as significant renovations like a gut rehabilitation. We're working with the American Society of Interior Designers to develop a set of best-practice guidelines for other types of renovations.

Q & A   (continued)

What steps can anyone take to upgrade his existing home, regardless of LEED certification? There are a multitude of facets for upgrading in terms of different energy measures, material selections, and mechanical and ventilation systems. However, one of the biggest hitters is to have an energy audit, because that can help to identify the low-hanging fruit for the homeowner. It's a good first step.

How quickly does a LEED home typically recoup any higher up-front building costs? The average cost to build is about $300,000, with the green features at about $8,500. When you amortize both of those over thirty years at 6 percent, and consider the offsets in utility savings, the net cost of ownership for the green features is pretty small. You can actually own a LEED home for a few cents more a day.

So green building isn't just for the wealthy? The program cuts across all residential sectors—custom, multifamily, low-rise, and affordable housing. What's exciting to us is that green building is feasible for the affordable housing sector, a portion of the marketplace that we've made a targeted effort to work with.

What about the resale value of green homes? Once the program is beyond its infancy, my guess is that we'll see people accruing gains through resale value because of LEED certification or green qualities.

Emily Mitchell is the former assistant program manager for the U.S. Green Building Council's program for LEED for Homes.

Save the Planet in Thirty Minutes or Less

🔆 reduces CO$_2$
💚 improves health
💲 saves money
🕐 saves time

- Do a mini energy audit for a few minutes after dinner each night this week. How many lights are on in your house? Do you leave your cell phone charger plugged in? Does the porch light stay on all night? How many electronics have standby lights? Improve your conservation techniques. 💲 🔆

- Replace the three most frequently used lightbulbs in your house with energy-saving compact fluorescent lightbulbs (CFLs). Buy an extra bulb and give it to a friend. 💲 🔆

- Depending on the season, turn your home's thermostat up or down two degrees. 💲 🔆

- Buy at least one power strip, which makes it easy to turn things off without unplugging them from the wall and helps reduce phantom power load. Devices known as "smart" power strips can intuitively stop idle current drawn from your outlets even without being shut off. If a device offers an option to turn off standby lights, use it. When your gadgets and tools are fully revived, detach their chargers from the wall. 💲 🔆

- Have trouble keeping your showers short and sweet? Change to a low-flow showerhead and take quicker showers. Opt for showers over baths and turn the water off while you shave, brush your teeth, and do the dishes. 💲 🔆 🕐

So You Want to Do More

🔆 reduces CO$_2$
💚 improves health
💲 saves money
🕐 saves time

- Get a home energy audit, which will help you assess how much energy your home uses and what measures you can take to make it more efficient. Do-it-yourself information is provided online at the Home Energy Saver website (hes.lbl.gov); hiring a professional to create a customized plan will help increase your home's efficiency even more. Some utility companies offer this service; the U.S. Department of Energy also offers good advice for hiring a pro. $ ⚡

- Make sure your home is properly insulated. Pay attention to hot spots like the attic, beneath floors, above unheated spaces, and around walls in a heated basement. $ ⚡

- If you have a hot water heater, buy a jacket for it. $ ⚡

- Don't plant your refrigerator near a stove, dishwasher, or heating vent. Move it to a cooler spot. Vacuuming the coils every few months eliminates dirt buildup and pumps up efficiency. $ ⚡

- If it makes sense, upgrade your windows. Otherwise, installing storm windows or plastic barriers is an affordable way to add insulating layers. $ ⚡

- If you have central air-conditioning, cleaning the evaporator and condenser coils and airflow components will keep the system working at higher efficiency. $ ⚡

- Buy green power. More than six hundred regulated utilities in thirty-plus states offer green pricing programs through which consumers pay a premium to support renewable-energy initiatives. The green power won't be delivered directly to your house, but you'll be making an investment in the future. If your local provider doesn't offer this service, consider buying renewable-energy certificates through green-e.org. The Green-e logo indicates that sellers are independently verified. Carbon offsets are also an option. ⚡

- Repair leaky faucets and toilets. A dripping faucet can waste twenty gallons of water a day. A leaking toilet can use two hundred gallons of water per day, according to the EPA. $

- Keep a watering can near your kitchen sink, and pour gray water into it. Think of it as water "scraps." Use it to water your houseplants. $

- If you have a choice between using natural gas or electricity to run your appliances, choose natural gas. Appliances that run on gas are generally cleaner in terms of CO_2 emissions and cost less to operate. 💲🔦

- Make a habit of washing laundry and dishes only with full loads. When washing dishes by hand, fill up the sink then turn off the water. 💲🔦

- Replace your showerheads with low-flow varieties, even if you rent. To check the flow rate of your shower, turn it on all the way and see how long it takes to fill a one-gallon plastic jug. Less than 24 seconds indicates more than 2.5 gallons per minute.[26] 💲🔦

- To further increase lighting efficiency, install motion sensors, dimmers, and timers where appropriate. 💲🔦

- For news about cutting edge home design and products, check out TreeHugger.com's design and architecture posts. At PlanetGreen.com get seasonal green home and garden advice. Want input from others about specific eco-techniques? Get involved in Tree Hugger forums (www.treehugger.com).

8

Week Six

Dressing Up:
Clothing and Personal Care

Your mission:
Green your style.

Like food, clothing is a cross section of farming, mechanical processing, transport, and retail. Which means not only that it uses a lot of energy, but that it has a lot to do with your eco-style. Luckily doing good doesn't mean sacrificing looking good. This week, it's time to overhaul both your closet and your beauty regimen—it's time to start wearing your heart on your sleeve.

COTTON AND OTHER TEXTILES

As one of the top five U.S. crops, cotton is no small player in the agricultural market. Sixty percent of it is transformed into apparel. Even more so than other crops, cotton is tough to grow—forcing farmers to throw enormous amounts of energy-intensive chemicals and fertilizers at it. To produce just one pair of regular cotton jeans takes three quarters of a pound of fertilizer and pesticide. Each T-shirt takes one third of a pound.[1]

Choosing clothing made from organically grown fibers is one

of the best ways to reduce the footprint associated with your wardrobe. Cotton is the largest crop produced for making natural-fiber textiles, but even organic cotton requires enormous amounts of water to produce and process. Hemp, bamboo, ramie, linen, and silk can also be organically produced. Hemp and bamboo, even when not organically grown, are still far better for the environment because they require little if any fertilizer to grow successfully. A very high yield crop, hemp grows well without herbicides, fungicides, or pesticides; captures CO_2 as it grows; and can be woven into sophisticated textiles with a very soft hand. The production of conventional cotton, on the other hand, consumes almost half of all agricultural chemicals used on American crops.

Wool, alpaca, and cashmere, especially organic styles, are also excellent choices. These, along with other animal-hair textiles, are considered renewable resources that are biodegradable and gentler to the environment than oil-based synthetics. Organic versions of the materials come from animals raised using methods that reduce or eliminate the need for agricultural chemicals that pollute soil, air, and water. Lyocell, commonly known by the brand name Tencel, is a textile made from wood pulp. It is recyclable and biodegradable and can be incinerated or digested in sewage without emitting any significant amount of carbon dioxide. Anything made of nylon, polyester, or acrylic, by the way, was originally derived from fossil fuel.

> Q: Is silk eco-friendly?
>
> A: The protein fiber spun by moth larvae, silk is a renewable resource and readily biodegradable. However, some silk farmers use chemicals to grow the trees on which moths feed, dyes can be an issue, and animal rights activists dislike that, generally, moths are killed in order to harvest their bounty. Eco-friendly silks are available, though the vast majority of commercial silk farming occurs in Asia, meaning the materials cannot typically be locally sourced.

The Thing with Bling

Jewelry is supposed to make you feel good, right? Spoiler alert: To mine for gold and silver, large-scale operations use cyanide and mercury, which, along with the other pollutants used, can devastate nearby ecosystems, especially waterways. Mining the gold for a wedding ring with 5 grams (less than 0.2 ounces) of the precious metal could leave behind more than 3 *tons* of waste.[2]

Precious gems, especially diamonds, can be associated with human rights abuses and are sometimes sold illicitly in war zones to fund rebels. For socially and environmentally just baubles, shop with care. Reputable dealers should be able either to provide authenticity certificates or to relay which company supplied their wares. (You'll have to do the research from there.)

The Kimberley Process Certification Scheme was established in 2002 to set trading standards. While it has drastically reduced the number of "conflict" or "blood" diamonds circulating globally, flawlessness is tough in an industry that's practically synonymous with illicit trade. Canada discovered a treasure trove in the Northwest Territories in the 1990s, and quickly carved out a large slice of the market with its conflict-free "polar bear" diamonds. (Australian diamonds are also considered conflict-free.) However, no diamond mining is especially gentle on the land, and some environmentalists fear that the mining will ecologically devastate Canada's boreal region.

What can you do in the face of all this potentially sketchy bling? The highly bedecked will be happy to know that Tiffany & Co. gets a nod for standing up against wilderness-threatening mining, supporting better operations and making an exemplary effort to eradicate blood ice from it's velvet-lined cases. Other jewelers, such as greenKarat, use recycled gold. Brilliant Earth's diamonds come exclusively from Canada, and the company swears a commitment to environmentally sound gold and platinum. Some environmentalists argue, however, that the only green diamonds are no diamonds at all. If you're on the fence, choose antique or estate jewelry; the reuse factor means you won't compound any of the issues.

Dyeing

Dyeing all types of textiles commonly includes using heavy metals and other toxins such as dioxin and formaldehyde, which can create environmental havoc. To flush conventional synthetic dyes from garments, enormous amounts of water must be used. Since a large percentage of textiles are produced in developing countries where pollution is not always well monitored, this wastewater can end up in rivers or other bodies of water. People with chemical sensitivities may find themselves bothered by dyes or other fabric finishes (such as formaldehyde, which is used to produce "non-iron" garments). Low-impact and natural dyes do exist, though they're not always impact-free; non-dyed, natural-color clothing is always a safe, if boring, option.

DRESSING ORGANICALLY

With the success of the organic food market, clothing manufacturers and retailers such as Patagonia, Nike, Levi's, and even Wal-Mart are starting to create commercial demand for organic textiles. In fact, demand for organic cotton far outweighs supply—only 6,577 acres of certified organic cotton were planted in the United States in 2005—that's less than 0.05 percent of cotton acreage overall. Even at that low rate, the United States, along with Turkey, is the world's largest producer of organic cotton.[3]

> Organic farming releases fewer CO_2 emissions into the air and uses
>
> **50%**
>
> less energy.

Though there's no government-sanctioned label for organic clothing as there is for organic food, many manufacturers mention when their clothing is made from organically produced materials. Also, unlike organic food, organic apparel

Dress Like a TreeHugger

Long gone are the days when dressing green meant swaddling yourself in shapeless, colorless hemp or, more extreme, going buck

THE GODDESS

Prana organic cotton and spandex tank top, Blue Canoe organic cotton and Lycra shirt, Global Girlfriend organic-cotton yoga pants, Simple leather and organic-cotton canvas sneakers with recycled-tire soles, aGaiN NYC recycled fabric yoga mat bag, Stewart+Brown hemp canvas duffel bag.

THE URBAN HIPPIE

iWood sunglasses with sustainably harvested wood frame, Green Label Organic cotton T-shirt with low-impact dyes, Loomstate organic cotton jeans, Voltaic messenger bag with solar panels, Worn Again sneakers made from 99 percent recycled materials.

naked. Today's green fashions are built to suit modern needs—
from the office to the runway to everyplace in between.

THE NINE-TO-FIVER

Boll Organic cotton dress shirt,
Nau organic cotton and spandex
pants, Tom Bihn molded cork
laptop case, Patagonia stitched
leather and latex loafer.

THE FASHIONISTA

Linda Loudermilk bamboo-wool
and Lyocell dress, Moonrise Jewelry
gold earrings with fair-trade gems,
Stewart+Brown Mongolian,
cashmere scarf, Ecoist clutch made
from recycled candy wrappers.

has no obvious and immediate personal health benefits, meaning that manufacturers may have a tougher time getting an industry-wide label approved. Some brands have introduced voluntary information. In 2006, for instance, Timberland added a "nutrition label" to its footwear, detailing how and where a product was manufactured and its impact on the environment.

SECONDHAND AND VINTAGE CLOTHES

Secondhand clothes help cut down on the waste associated with manufacturing, buying, and selling new clothes. From the threads at the Salvation Army to the tailored pieces made from recycled garments by Junky Styling—beloved by celebs such as Gwen Stefani—there are some serious finds among previously worn garments and accessories regardless of your budget. The other side of the equation—donating—saves resources and prevents landfill rot. Have jeans that you'd rather never part with? For a price, the magicians at Denim Therapy can rethread your tattered, thinning jeans (no patches here), keeping your fave pair in working order, and your wardrobe a little greener.

CASE STUDY: The Swaporamarama

One part recycling center, one part fashion event, the Swaporama-rama is the Super Bowl of stylish reuse. For ten bucks and a bag of used clothes, "shoppers" pick through secondhand items then transform their finds into new getups at sew-it-yourself stations or with the help of on-hand experts. Since their inception in a tiny New York apartment, dozens of open-to-the-public seasonal swaps have sprouted up all over the country. Each event recycles about seven thousand pounds of clothing, thereby diverting it from the landfill, encouraging reuse instead of consumerism, and extending a sense of DIY enthusiasm among people and communities.

Action: Annual Savings in pounds of CO_2 • Donate all of your unwanted clothing

CLEANER LAUNDRY

Slightly tweaking the way you use your clothes dryer—or better, not using it at all—is a wise and simple way to decrease the amount of energy you use at home. To increase the efficiency of yours—and thereby save carbon emissions—place it in an area that's typically warm. Clean the lint filter after each load, and turn it on only when it's full. If your dryer features a moisture-sensor option, use it; this ensures the machine automatically shuts off when the clothes are dry. Better yet, the most ecologically sound thing to do is line-dry your clothes whenever possible, which uses no energy at all. Apartment dwellers: try a space-efficient rack.

Using the cold-water setting on your washing machine saves 85 percent of the money and energy consumed. If your washer has spin options, set it to a high- or extended-spin setting, which will ring clothes out as much as possible before you put them in the dryer. Front-loading washers are more efficient because they use less water and energy and are gentler on your clothes.

Synthetic detergents, like many other conventional cleaning products, contain surfactants, made from petrochemicals, which pollute the water outside and can leave residues on clothes. Non-toxic, readily biodegradable detergents are a better option. Another refresher: Look for alternatives to chlorine bleach, which can damage fabrics and release harmful toxins into the water stream. Swap out the fabric softener for a half cup of vinegar instead. Optical brighteners are another eco no-no, as these products coat clothes with a chemical that reflects light, making clothes appear whiter or brighter. Many are derived from benzene, a known carcinogen and highly toxic compound. They don't readily break down, are toxic to fish, and can create bacterial mutations.[4]

instead of throwing it out: 165 • Wash your clothes in cold water: 327

The Dirt on Dry Cleaning

Because most traditional dry cleaners use the chemical per-chloroethylene, or "perc," they may not leave your clothes as clean as you think. Perc is what gives your dry cleaning that weird smell. It's harmful to the environment and your health, can be stored in the body, and has been linked to increased risks of various cancers; reduced fertility; eye, nose, throat, and skin irritations; and smog. If your clothes have a strong chemical scent when you get them back, they are likely still damp. Ask the cleaner to properly dry them out thoroughly. Particularly dangerous for those working with it, perchloroethylene can also escape from dry cleaning shops into the outdoor air and well water and can seep into the air of your home if not completely removed from clothes during the cleaning process.

Improving existing equipment is one strategy for reducing the amount of perc that escapes. Better options are also slowly beginning to take shape. Wet cleaning uses water as the main solvent in specialized machines. The liquid carbon dioxide method, which uses CO_2 captured from the production of industrial chemicals and natural sources, is also a better alternative. Both methods are recommended by the EPA. A slightly controversial third, Green-Earth cleaning, uses silicone-based solvent in specialized machines. Unfortunately, the high costs associated with the equipment and materials necessary to implement these new cleaning methods has limited their introduction.

LOOKING GOOD: NOT ALWAYS PRETTY

Don't you just hate it when an expensive new face cream makes you break out in a rash? Ever wonder if it was the PEG-80 sorbitan laurate, drometrizole trisiloxane, or the oxybenzone that messed with your complexion? It turns out it's more than just money we throw down the drain when we toss out criminal lotions. Ironically, many products that claim to be making us more beautiful and healthy are actually laden with chemicals that aren't so great for our bodies or the planet. With the exception of color additives, the U.S. Food and Drug Administration does not review or regulate personal care products—not even those made for kids—or their ingredients for safety before they hit store shelves.

One out of every one hundred personal care products on the market contains known or probable carcinogens. With industrial chemicals as their basic ingredients, grooming products also contain pesticides, reproductive toxins, endocrine disruptors, plasticizers, degreasers, and surfactants. Of the 10,500-plus ingredients in these products, nearly 90 percent do not have publicly available safety reports, according to the Environmental Working Group.

Meanwhile, the average person uses close to ten grooming products each day—shampoo, toothpaste, soap, deodorant, hair conditioner, lip balm, sunscreen, body lotion, shaving products, and cosmetics that expose us to approximately 126 unique ingredients that subsequently seep through the skin and rinse down the drain.[5] Most products are presumed to be safe, and many are. But no one knows the cumulative effects of using these products over a lifetime. For

Eco-myth:
Dressing green means wearing mostly organic cotton and hemp.

Fact:
Sustainable materials are a great choice, but you can significantly reduce your closet's environmental impact by caring for conventional clothes efficiently. For example, washing a cotton T-shirt—whether or not it's organic—in cool water can reduce its overall impact, including manufacture, by about 10 percent; steering clear of the dryer cuts the garment's total CO_2 impact by 50 percent.

• Run the washing machine only with a full load: 99 (Sources: Rocky Mountain

Beauty Inside Out

Reading ingredients labels on what seem like even the simplest of products can be like trying to translate a foreign language. And many chemical names represent large groups of differing chemicals, some of which may be benign while others in the same group are known or suspected to negatively impact human health. And marketing lan-

INGREDIENT	CODE NAME	LURKS IN
Coal tar	FD&C Blue 1, Green 3, Yellow 5 & 6, D&C Red 33	Hair dye, makeup, some strong scalp-treatment shampoos
Formaldehyde	Formaldehyde, formalin, DMDM Hydantoin	Nail treatments, blush, face powder, and other cosmetics
Fragrance	Fragrance, *parfum* (these can be made up of hundreds of unlisted chemicals)	Lotions, shampoos, soaps, cleansers, moisturizers, makeup, perfumes, etc.
Mercury	Thimerosal	Pain and wound treatments, eyedrops, occasionally eye makeup such as mascara (the FDA has banned mercury in all non-eye-area cosmetics)
Lead	Lead acetate	Some hair dye for men (many hair dyes also contain harmful ammonia)
Parabens	Methyl-, ethyl, butyl-, propyl-, iso-, and butyl-parabens	Facial cleansers, liquid soaps, hair conditioner, toothpaste, shaving cream
Phthalates	Dibutylphthalate, DPB, DEHP, DHP. Not listed when used in fragrance	Perfume, nail polish, hair spray, soap, shampoo, mascara, etc.

Institute, EU Carbon Calculator)

guage and terms such as "natural," "hypo-allergenic," and "organic" can be misleading. Below, we decipher some of the most common chemicals lurking in personal care products. Note, however, that their presence in products you may use is cause for precaution, not alarm.

PURPOSE	NOT SO PRETTY	ALTERNATIVE
Color agent or additive; dry scalp treatment	Some ingredients are known carcinogens, others may contain carcinogenic impurities	Skip the hair dye or use it less frequently, or use henna, a plant-based dye; milder dandruff shampoos; natural makeup products
Preservative	Very likely a carcinogen, allergen	Products that use vitamins E or C or citric acid
Added or cover-up scent	Very common allergic reactions, including contact dermatitis; respiratory irritant	"Fragrance-free" products (check the ingredients), or those scented with essential oils
Preservative	Neurotoxin	Products that don't use it
Coloring agent	Neurotoxin	Use henna or stay a silver fox
Used as preservative to prevent clumping and bacterial growth	Endocrine disruptor, possible carcinogen	Products that use vitamin E (tocopherol), vitamin C (ascorbic acid), or citric acid
Makes materials more pliable, prevents cracking, retains scents by "fixing" perfumes to reduce evaporation	May cause developmental and reproductive problems	All-natural soaps and shampoos, organic products, products scented with essential oils

Beauty Inside Out (continued)

INGREDIENT	CODE NAME	LURKS IN
Triclosan	Triclosan	Lipstick, lip-gloss, antiperspirant and deodorant, cleansers, acne treatment, moisturizers
Toluene	Toluene, phenylmethane, toluol, methylbenzene	Nail polish and nail polish hardeners and removers
Sodium lauryl sulfate	Sodium lauryl sulfate, SLS, sodium dodecyl sulfate, SDS	Soaps, shampoos, toothpaste, other sudsy products

SOURCES: U.S. FDA, Environmental Working Group, Consumer Reports Greener Choices

this reason, we apply the precautionary principle when it comes to grooming and beauty.

Multiple studies have shown that common water pollutants—many from personal care products and pharmaceuticals—are disrupting the hormone systems of wildlife. Some show that estrogen from birth control pills can affect the reproductive systems of salmon and other fish. Another found a relationship between phthalates and the feminization of male infants in the United States, and named a common fragrance fixative, diethyl phthalate, as a possible culprit. Parabens—used as antimicrobial preservatives—were recently found in human breast tumor tis-

PURPOSE	NOT SO PRETTY	ALTERNATIVE
Antibacterial	Skin and eye irritant, possible antibacterial resistance buildup, probable human carcinogen	Products that use grapefruit seed extract or vitamin C (ascorbic acid) instead, natural deodorants, beeswax-based lip balm
Solvent	Neurotoxin, affects kidneys, may cause developmental defects	Buff instead of polish or use products containing water-soluble ingredients or that have eliminated toluene
Suds-forming detergent and surfactant	Skin irritant, especially for those with eczema, psoriasis, or other dry-skin conditions	Look for vegetable oil–based soaps, products that are "sodium lauryl sulfate–free," natural toothpastes, etc.

sue, leading researchers to question a link with deodorant.[6] While enough research has not been done to determine conclusively how severely these products are affecting us, such studies are cause for concern.

Safer Sunscreen

Everyone knows that sun protection is essential to staying healthy while spending time outdoors. Sunscreens and sunblocks with high sun protection factor may have put a dent in the market for baby oil and tinfoil, but what many people don't realize is that the higher the SPF, the more chemicals a lotion contains.

For the Ladies

In 1985, the super-absorbent polymer polyacrylate was banned from tampons due to issues that linked it to toxic shock syndrome. While safer today, most conventional feminine care products are made with fibers that have been bleached with chlorine, which produces dioxin, which can then be released into the environment and our bodies. Since dioxin settles in fat cells, it bioaccumulates over time. To reduce exposure to this and other pollutants, use unscented, unbleached cotton tampons and sanitary pads. (You might consider going organic, too; some believe that pesticides and herbicides also could taint these products.) Better yet, for those without sensitivity to latex reusable menstrual cups made from latex are generally safe and reduce landfill waste. For the eco-brave, reusable pads are also available. Tampons, by the way, have a slight life-cycle advantage over disposable pads, due to the plastic components found in most conventional versions of the latter.[7]

Sunscreen undoubtedly reduces the risk of squamous cell carcinoma, the second most common form of skin cancer. But its effectiveness against basal cell carcinoma—the most common type—and malignant melanoma is uncertain. One study at the Queensland Institute of Medical Research, in Australia, found that sunscreen use reduced the risk of developing squamous cell carcinoma by 40 percent but did not reduce the risk of developing melanoma or basal cell carcinoma.[8]

Number of chemicals banned from cosmetics:

EU: 1,100
U.S.: 10

A comprehensive Environmental Working Group (EWG) study found that 716 of 853 sunscreens offered inadequate sun protection or contained ingredients with significant safety concerns. At least 50 percent made misleading claims, such

Well-groomed and Green

These TreeHugger-approved personal care products are beautiful on the inside to keep you that way on the outside.

Top Five for He-Huggers

1. Aubrey Organics Men's Stock Natural Dry Herbal Pine Deodorant
2. Avalon Organics Aloe Unscented Moisturizing Cream Shave
3. Nature's Gate Crème de Mint Natural Toothpaste
4. John Masters Organics Zinc & Sage Shampoo with Conditioner
5. Lavera Men Care Face Wash

Top Five for She-Huggers

1. Pangea Organics Facial Toner
2. Sothys Secrets de Sothys Body Smoothing Balm
3. Honeybee Gardens WaterColors Nail Polish and Odorless Polish Remover
4. Dr. Hauschka Skin Care Intermezzo Mascara
5. Rich Hippie Bliss Organic Perfume

Top Five for Wee-Huggers

1. Earth Mama Angel Baby Angel Baby Oil
2. Badger Diaper Cream
3. Seventh Generation Baby Wipes
4. Dr. Bronner's Baby Mild Organic Liquid Soap
5. California Baby No Fragrance SPF 30+ Sunscreen Lotion

as "all-day protection" and "blocks all harmful rays," and 12 percent of sunscreens with SPF 30 or higher protected only against UVB radiation, which causes sunburn, but offered no protection against UVA rays, the ones linked to skin damage and aging. Though there's presently no standardized rating for UVA protection as there is for UVB, we will likely see one soon.[9]

Suspect ingredients in sunscreens include phthalates; diethanolamine (often listed as DEA), an emulsifier and suspected carcinogen; and the ultraviolet blocker benzophenone (also called oxybenzone), which is known to be absorbed through the skin and is a known allergen and suspected endocrine disruptor. Some contain parabens, which are endocrine disruptors and possible carcinogens, and PABA (para-aminobenzoic acid), an allergen and possible carcinogen. Many brands are now paraben- and PABA-free, and usually say so on their labels.

Paradoxically, the EWG report found, many ingredients also break down in the sun. And when we swim or bathe, many of these chemicals can wash off our bodies and into the water system, where they affect marine wildlife. Nanoparticles—microscopic particles measured in billionths of a meter—also raise some concern. While micronized and nano-scale zinc oxide and titanium dioxide provide good UVA protection with little health concern, EWG notes that nanoparticles may contain toxic properties of which we are not yet aware; their tiny size means they may seep into our bodies more readily.

Wearing sunscreen is crucial in many situations and should absolutely be practiced. Covering up and staying out of the sun when possible are also good options. After three decades of debate, the FDA is finally getting to set mandatory standards for sun-protection products, which should lead to better quality control and a consistent benchmark for UVA ratings. EWG's cosmetic safety database, Skin Deep, provides a current and extensive list of ratings and recommended products.

A Brief Guide to Greener Sex

1. Get hooked.

Just haven't been able to find that special someone? Quit whining. The green community's answer to J-Date, matchmaking website Green Singles, is calling. Or join the communities at Planet Green.com or Tree Hugger.com and mix it up.

A Brief Guide to Greener Sex (continued)

2. Linger in your lingerie.

Think timeless, not trashy, to keep unmentionables in circulation and out of the landfill. (After all, nobody wants a secondhand thong, no matter how cute it is.) Sexy new organic cotton, hemp, silk, and bamboo offerings are not your mama's knickers.

3. Slip and slide.

When it comes to lubrication, avoid petroleum products, artificial scents, flavors, and colors. Step it up and go organic.

4. Better booty.

Toys made from softer plastics and PVC can contain phthalates, which we simply cannot recommend using you know where. Opt for glass, metal, silicone, hard plastic, and elastomer options. Two more words: Rechargeable batteries.

5. Keep the numbers down.

Nothing says "I love you" like population control. All latex condoms may or may not be biodegradable, but at least they're not made of polyurethane. Lambskin (yes, they still make those) is all natural, but gets a big zero for STD protection.

Q & A

Eco-model and Activist Summer Rayne Oakes, on JLo, Cute Undies, and the Greenest Outfit of All

What is eco-fashion? Design that considers environmental and social responsibility throughout the entire process of making, using, and reusing a garment.

You're an eco-conscious model and environmental advocate. You're also an Ivy League grad with some serious knowledge about sewage sludge. Why did you pick the fashion industry as your platform for spreading the green

Q & A (continued)

word? Don't tell me you *don't* look to see what JLo is wearing on the red carpet. Fashion is glamorous, transcends language, and attracts people who wouldn't otherwise be interested in environmental issues. That makes it the ultimate communication tool.

Our closets aren't the first place most of us consider when contemplating our impact on the planet. Why should eco-savvy consumers care about their clothes? The textile industry supports economies worldwide. How can we use the business as a tool for sustainable development? Think of your purchases that way, and you're guaranteed to understand that you *can* make a difference.

What's the easiest way to reduce the impact of your closet? Thinking about how you care for your clothes.

What are the most difficult aspects of dressing green? Accessibility and choice.

How does your closet stack up? I buy from local eco-designers, choose vintage, and take the old stuff to thrift shops and donation centers. Right now, I'm holding on to a bunch of bulky T-shirts that I'm never going to wear—I'm just waiting for the right designer to turn them into cute undies.

Any guilty pleasures? I'm not a shopaholic, so I would have to say that my one guilty pleasure is traveling.

Sustainable living doesn't have to mean sacrificing your aesthetics, but "eco-friendly style" often gets equated with "earthy looks." Do green designers and models have an added responsibility to create appealing objects of de-

Q & A (continued)

sire? Yes! If you are a visible leader, it is important to wear your values.

How would we dress in a perfect green world? It's a tough question. I'd like to go naked, but that's probably not going to happen. In a greener world, clothes would be made locally with sustainable materials in fair conditions and they'd be very durable. We would strive to create clothes that we cherish and that help people and the planet.

Summer Rayne Oakes is a model, writer, and sustainable-business strategist.

Save the Planet in Thirty Minutes or Less

reduces CO_2

improves health

saves money

saves time

- Whenever possible, wash your clothes in cold water only, which reduces environmental impact, and use ecologically safe detergents. Skip using the dryer whenever you can, which can reduce the overall climate impact of a garment by about 50 percent, according to a life-cycle assessment study by the Cambridge University Institute of Manufacturing.[11] Skip ironing if you can.

- Tally up the number of personal care products you use every day and every week. Surprising, isn't it? Begin to cut back on their number or at least how often you use them, and replace one daily or weekly item with an eco-friendly grooming product.

- Read the label on your shampoo bottle or other product. Are there ingredients you don't recognize or approve of? Use the Environmental Working Group's Skin Deep cosmetic database (www.cosmeticdatabase.com) to look up ingredients not listed on the table on page 156–9.

- Pledge to buy one pair fewer shoes this year. Donate the money you save to an environmental cause.

- Reduce trips to the dry cleaner by washing clothes at home, extending time between cleanings, and buying fewer clothes that require dry cleaning.

So You Want to Do More

reduces CO_2
improves health
saves money
saves time

- Look for garments that use recycled content. Research shows that the environmental impact of recycling worn-out polyester into new polyester fiber, for instance, is significantly lower than making that same fiber from virgin materials.

- Donate unwanted clothing instead of throwing it out, or hold a clothing-exchange party with friends. Recycling or donating your clothes helps reduce CO_2 emissions associated with incineration or rotting in a landfill. Often you can get a tax deduction for donating.

- Buy vintage, which reduces the need for more primary production.

- Extend the life of your clothes. Repair them instead of getting rid of them the moment a button falls off. Choose quality over quantity. Buying things you'll wear for a long time saves energy and keeps landfill volume down.

- Choose apparel made from hemp and bamboo. Even nonorganic varieties of these materials are more sustainable, since they require little or no fertilizer to grow, thereby producing far fewer CO_2 emissions. Organic cotton, wool, and other eco-friendly materials are good choices, too.

- Cows create loads of greenhouse gases. Could you buy fewer shoes made from leather? We're not asking you to stoop down to vinyl, but why not give canvas and hemp a chance?

- Skip the conventional perfumes as much as possible. Read the label: Avoid products with the words "fragrance" or "parfum," which indicate the presence of phthalates.

- Durability is an important aspect of any eco-wardrobe; avoiding trendy throwaway clothes is key. Still, style is our thing. That's why TreeHugger.com highlights hot-yet-timeless fashions for every budget each season. Get the latest beauty hints and grooming tips at PlanetGreen.com.

9

Week Seven
Getting to Work:
Building a Better Office

Your mission:
Reduce the carbon cost of doing business.

In an ideal world, we'd all work for companies that use Energy Star appliances, offer bicycle parking, reimburse employees for carpooling and purchasing hybrid cars, and have solar panels on the roof. But even if in reality you work for some mega-corporate conglomerate that occasionally makes headlines for dumping millions of gallons of oil on penguins (hey, everyone has to make a living), it doesn't mean you have to check your values at the door when you punch the time clock.

This week focuses on how you can make a difference at work, from commuting to stocking a break area to choosing the best fax machine. Whether you pay the bills by putting your time in on the floor of the New York Stock Exchange, teaching elementary school, hawking action figures on eBay, or serving up cocktails, there's something all of us can do.

Conducting a Waste Audit

Whether you work in a cubical farm or out of the spare bedroom, you can streamline business operations, save money, and do good by the environment by conducting a waste audit. Begin your audit by separating waste into categories and taking visual stock of what you're tossing out and recycling and in what quantities—paper, printer cartridges, aluminum cans, glass, plastics, packing materials, and so forth—for one week. This type of granular breakdown can help you identify opportunities to reduce materials used (say, by making double-sided copies or printing in draft mode), recover others for reuse (such as packaging), and see what more you can recycle.

If your audit is a success, consider enacting it company-wide. It may be useful to examine purchasing records and to speak with other employees about business processes and activities that could be made more efficient and less trash-intensive. Waste audits can be very effective at convincing corporate higher-ups that going green will save them money. Most companies pay to have their trash hauled away, but they can earn dollars back for recycling paper, containers, toner cartridges, corrugated cardboard, and more.

The Paper Chase

Isn't it weird that in the digital age we still consume enormous amounts of mashed-up, bleached tree pulp, most of which gets used just once then tossed? Paper manufacturing is one of America's most energy-intensive industries, releasing about 35 million tons of CO_2 into the atmosphere each year.[1] Using virgin wood to make paper helps deforest the planet, which both reduces the number of trees available for sinking CO_2 and releases carbon contained in the felled trees. The average office worker throws out about 350 pounds of paper per year, according to the Natural Resources Defense Council, compounding the problem. Save paper—and CO_2

emissions—by thinking twice about what you print out, making double-sided copies, and using scrap paper to take notes or print drafts. Choosing paper with high post-consumer recycled content—which means that materials were recovered from paper that was previously used by consumers—means resources are reused; it diverts waste from landfills and creates demand for recycled products. Products marked with "post-industrial recycled content" are less beneficial, since the term refers to waste such as mill scraps that never reached consumers and that manufacturers already reuse in order to save money.

The Well-Equipped Office

If you work for yourself or you work at home, you probably have control over which office supplies you choose. Raise your eco-credit—and save money—with energy-saving appliances, post-consumer recycled paper, recycling bins for paper and cardboard, and reusable necessities such as flatware and plates in the break room. Choose energy-saving fax machines, copiers, scanners, and printers, for example, which use about half as much electricity as standard equipment and also default to a low-power sleep mode.

If you work in a place where someone else calls the shots, lobby your company, colleagues, and facilities managers to do the same. Suggest that your company install motion sensors for lighting and purchase its power from renewable sources.

Greener Computing

The average desktop computer consumes between 80 and 250 watts; with laptops, those numbers dip to 30 to 80. With computers around the world on day and night, it requires enormous amounts of energy to keep us all connected. And while most screens in the United States are modern LCDs, numerous energy-sucking CRT (cathode-ray-tube) monitors—which use at least twice as much energy as LCD screens—are still in place all around the world. With computer life spans shrinking as we load them up, wear them out, and turn to ever-faster processing power, we create a hulking heap of obsolete computers every year.

Tips for Buying a Greener Computer

1. Don't buy one. Instead try to get one for free at Craigslist or Freecycle or from friends.

2. If you have to buy, search EPEAT to find a machine with the highest level of environmental sensitivity based on your specifications.

3. Consider purchasing a used or refurbished model. About 98 percent of the total environmental cost of computing is in the manufacture of the product. One caveat: Older machines could potentially use a lot more energy.

4. Once you get your machine, run on corded electricity as often as possible. Batteries are made of nasty stuff and are energy hogs. Using the grid extends battery life, which is also more economical.

To consume less power and make your machine last longer, use "lighter" programs—that is, software that takes up less hard drive space and uses less system memory. Whenever possible, use programs that use less processing power to do the same work; for example, using TextEdit instead of Microsoft Word will create less stress. Likewise, configuring your computer to view simple text Web pages and block more intensive applications—such as flash animation, which is often used to display ads—will help as well. As a rule of thumb, if the machine is working harder, it's probably using more energy.

To keep your computer functioning at maximum efficiency, adjust the power options in your control panel to energy-saving settings. Screen savers may seem low-effort compared to the amazing things we task our computers with each day, but they consume as much electricity as when your machine is idle, not sleeping. Bag the mesmerizing rainbow fractals and tell your computer to go into sleep mode, which reduces energy consumption by up to 70 percent. If everyone did this, the EPA estimates, we could save enough electricity each year to power Vermont, New Hampshire, and Maine; cut electric bills by $2 billion; and reduce carbon dioxide emissions by the equivalent of 5 million cars.[2]

Shutting Down. Nearly half of all corporate PCs in the United States are left on at night, costing businesses $1.72 billion in energy and emitting more than 14 million tons of carbon dioxide, according to a report by PC Energy Awareness.[3] When you're ready to throw in the towel, power down. It's a myth that leaving your computer on overnight is more efficient than rebooting it the next morning. Shutting down will conserve energy, reduce mechanical stress, and prolong your computer's useful life.

If your IT department requires you to leave machines on at night, ask if they could run updates and security checks on specific nights each week. Plug hardware into a power strip to shut down your whole desktop setup at once and ensure that pesky standby power isn't drawing

Eco-myth:
Shutting down your computer at night and rebooting in the morning uses more energy than leaving it on all night.

Fact:
Shutting down conserves energy, reduces mechanical stress, and prolongs the useful life of a machine.

a phantom load. Printers, scanners, and other peripherals that are used only occasionally can be unplugged until they're needed.

Getting the Lead Out. When it comes time to purchase a new computer, look for a machine that meets EPEAT (Electronic Product Environmental Assessment Tool) standards. Responsible computer and electronics makers should pay close attention to environmental impacts associated with manufacturing, packaging design, energy efficiency, and recycling. The comprehensive EPEAT database (www.epeat.net) allows you to search by criteria, and rates products according to their environmental performance.

Most computers today meet the European Union's fairly strict RoHS (Restriction of Hazardous Substances) requirements; make sure your new purchase does, too. Also, look for take-back programs—including both the machine you're done with and the one you're buying—and power consumption levels below the industry average. Toshiba and Dell have taken the lead, with a handful of notebooks earning EPEAT's Gold status, the highest level, based on several criteria such as materials selection, end-of-life design, product longevity, energy conservation, and packaging.

> **$3.5 billion:**
> Amount Energy Star–rated home office equipment saved Americans in energy costs in 2003.

Green computing isn't as elusive as it may seem. Many companies, such as Dell and HP, are starting to make PCs that not only are Energy Star–compliant but also contain reduced amounts of cadmium, lead, and mercury. Apple has significantly reduced a number of hazardous substances in its wares and plans to completely phase out polyvinyl chloride, brominated flame retardants, and arsenic from all of its products by the end of 2008. The company also takes back computers for free, regardless of brand, with the purchase of a new computer, and claims to never ship e-waste outside of the United States.

DOES YOUR BUILDING NEED A SICK DAY?

Indoor air quality problems are not limited to homes. In commercial settings and schools, poor indoor air quality can affect employee and student health and performance. Furniture, carpeting, paints, adhesives, insulation, and even copying machines can adversely affect indoor environments, as can radon, carbon monoxide, mold, and allergens. Large amounts of carbon dioxide—produced when people breathe—can also accumulate if spaces aren't properly ventilated, causing drowsiness, headaches, and decreased activity.

Legionnaires' disease, asthma, hypersensitivity pneumonitis, and humidifier fever have been traced to specific building problems. Other times, building occupants experience symptoms that are difficult to link to a specific source. This phenomenon, known as sick building syndrome, can result in dry or burning mucous membranes, sneezing, stuffy or runny nose, lack of energy, headache, dizziness, nausea, irritability, and forgetfulness. Proper design, operation, and maintenance of a building's ventilation system are essential to keeping employees healthy, which reduces the number of sick days taken, increases productivity, and therefore benefits the bottom line.

39:
Percentage of the municipal solid waste stream made up of paper products.

People generally have less control over their work environments than their homes, so if you suspect indoor air pollution is creating problems at your workplace, talk to your co-workers and supervisors to learn if others are experiencing problems. Discuss persistent symptoms with your doctor. If the problems are not easily solved, EPA regional contacts can help.

TELECOMMUTING

Video conferencing, instant messaging, online chat forums, and other tools make effective telecommuting and business travel reduction a real possibility. If you can work from home or take classes online, you can save precious CO_2 emissions from the trips to and from work. Consolidated work weeks—putting in four ten-hour days instead of five eight-hour days, for example—can also help cut energy costs associated with work travel. Alternatively, carpooling, taking public transit, biking, or walking to work are also good choices. Some companies will even agree to set up satellite offices closer to employees' homes.

GO GREEN, GET RICH

So you've started carpooling, you shut down your computer at night, and your office recycling program is in full swing. When payday rolls around, what will you do with your hard-earned greenbacks to continue closing the loop? Investing green can mean supporting sustainable companies, the environment, and your own portfolio. There's no single way to go about it; the method that works for you will depend on your personal goals, ethics, and research. It's beyond the scope of this book to dole out financial advice, but the information listed here surveys the basic framework of ideas behind green investing.

Buy Direct

The most obvious investment strategy is to put your money into a company focusing on green innovations, such as renewable energy technology (sometimes referred to as "cleantech" or "greentech"), or one that's shifting to sustainable processes—say, a car manufacturer that's overhauling its factories and products. Purchasing stock can help a company's value grow, which in turn makes it more valuable to other investors. Buying the stock can encourage others to do the same, raise the share price, and focus more attention on the company. Divesting money can send the opposite message—that you no longer agree with a company's mission or tactics. If that's how you feel, be sure to voice the reason why you dropped the stock.

Alternatively, some people hold the view that their money can do the most good when donated directly to a pet cause. You might consider not investing green at all, but rather putting your money where you expect the biggest return. In the end, that bigger chunk of change could allow you to put more dough toward the specific environmental concerns you really care about—we'll assume you'll use appropriate means to justify that end.

Mutual Appreciation

During the past thirty or so years, the strategy of socially responsible investing has grown into a multitrillion-dollar business. These mutual funds generally avoid investing in companies that support or produce alcohol, gambling, arms or weapons, or businesses that harm the environment or people. Increasingly, funds that place priority on environmental issues are beginning to emerge. But just because a mutual fund is labeled socially or ecologically responsible doesn't mean that it will match *your* personal values or invest exclusively in companies that line up with your moral compass.

(Sources: EU Carbon Calculator, Energy Star, StopGlobalWarming.org, Rocky

Some buy into only those companies that address specific environmental technologies such as renewable energy or recycling. Others emphasize a record of positive environmental performance.

Because a mutual fund is a product that needs to return a profit, individual companies within the collection must grow. Those that fail to make the grade are given the boot. The goal should be to find the sweet spot between maximizing financial return and maximizing environmental responsibility. Where that spot lies can be defined only by you.

Since socially responsible investors are driven by morals as well as money, most want to know that the actions, products, and values of the particular companies in a fund match their expectations. For this reason, it's wise to do your own research into how a fund is managed, how the companies are picked, and what exactly they do. How the fund manager screens for included companies is a good place to start. Does he use strict criteria—say, preferring companies that actively seek to improve their environmental records? Or does he cast a wider net, considering any company that has simply never been charged with an environmental infraction? These parameters—combined with desired return on investment—may help you clarify your personal criteria.

Shareholder Advocacy

Shareholder advocacy—letter writing, proxy voting, and creating resolutions in an effort to influence corporate behavior—is probably the most esoteric of the green investing paths. If you own stock in a company, you have the right to reinforce moves you think are wise and speak out against those you wish to discourage. Positive reinforcement is just as important—if you approve of recent decisions, let it be known that you'd like to see more of the same.

Admittedly, being heard can be complicated, but it is technically feasible. You could show up at the annual meeting, or use Internet forums to reach a critical mass, create consensus among other stockholders who feel the same way as you, and steer management on a course that you believe will improve the company and the environment. Be aware, though, that individual shareholders might not outweigh a corporate owner.

THE TRIPLE BOTTOM LINE

Using the three tenets of economy, ecology, and equity, the economic philosophy known as the triple bottom line suggests that companies will do best in the long run if they base their accounting on not just cold hard cash but the overlap of profit, the environment, and society. Sometimes this is referred to as "people, profit, planet," or "integrated bottom line" accounting.

Business as usual and our collective global footprint are putting our natural resources at risk. Meanwhile, human labor is often treated with disrespect. The triple bottom line seeks to retain a democratic, capitalist system, but one that doesn't worship making a monetary profit above all else. That is, it assigns natural capital and human resources value, and incorporates their overall costs and benefits into the bottom line. This requires a shift in understanding that business—indeed sustaining the entire global economy—is dependent upon the environment and the availability of natural capital, and so endeavors to benefit the environment, or at least do no harm to it. Conditions such as fair trade, fair working conditions, proper health care, and giving back to communities also contribute to the triple bottom line. To drive the point home, triple-bottom-line companies often refer to their investors as "stakeholders" rather than "shareholders," implying that they play a role that goes beyond just financial investment.

Though quantifying this bottom line can be difficult, the benefit isn't just about a cleaner planet and happier societies; it's also about dollars and cents. Environmental sustainability—conserving energy, reducing manufacturing waste and toxic runoff, creating cradle-to-cradle products, and developing materials-recovery programs—generally presents a more profitable course in the long run, despite arguments to the contrary or higher up-front implementation costs.

CASE STUDY: Interface

When Ray Anderson founded the Interface carpet company in 1973, his vision of environmental care went no further than complying with regulation. Twenty-one years later, he read Paul Hawken's *The Ecology of Commerce*, a treatise on how the modern rate of material use and consumption is endangering prosperity. Sustainable business—let alone the carpet industry—hasn't been the same since.

Anderson instantly decided that Interface—which owns brands including Bentley Prince Street, Flor, and InterfaceFABRIC—was on the wrong path. After analyzing every aspect of material flows, he set an ambitious mission: to be the first company to become completely zero-impact by 2020.

To reach that goal, Anderson laid out a sevenfold path that starts with eliminating all forms of waste from every aspect of his business and ends with redesigning commerce: Interface will become an exemplary new business model that demonstrates the intrinsic value of sustainability-based business. Today, every creative, manufacturing, and building decision is made with the intention of eliminating any negative impact to the environment and addressing the needs of society.

Though the company still has a long way to go to reach its ultimate goal, it has already developed industrial systems that decrease costs and reduce burdens on living systems. Implementing change ranges from minute efforts (an $8.50 nozzle that reduces water flow is saving 2 million gallons of water and $10,000 per

year at a Maine facility) to sweeping changes (seven facilities operate entirely on renewable electricity).

Another undertaking includes eliminating toxic substances
from products, vehicles, and facilities. To close the manufacturing
loop—another precept—Interface is constantly redesigning products and processes to reuse recovered materials. By harvesting
and recycling carpet and other technical products, the company
estimates that despite its reliance on petroleum-based products,
soon it may never have to use a new drop of oil again. Finally, it has
made a commitment to transporting people and products efficiently, and to educating people about the principles of sustainability and improving their lives and livelihoods.

Even before Interface made the commitment to go sustainable,
it had long been associated with a commitment to high-quality design and innovation. While it hasn't reached its goal just yet, its
processes and technologies have improved every year. In doing so,
the company has increased profitability and drastically decreased
it's environmental impact, proving that sustainability and the triple
bottom line can be built into both business decisions and extraordinary business models.

Q & A

**Joel Makower of GreenBiz.com on Corporate
Responsibility, Carbon Caps, and Why the Greenest
Jobs Have Nothing to Do with the Environment**

Can business sufficiently lead the way on environmental
responsibility in the absence of policy change and legislation? Business can and is leading the way, not necessarily because it is the "right thing to do" but because it makes good
business sense. Many companies are rooting out waste and
inefficiency in operations, buildings, product designs, and

Q & A (continued)

supply chains. This can reduce risk and liability as well as operational costs. But companies are also finding other value in this, including increased sales, quality, the ability to attract and retain talent and customers, and better reputations.

There's a trend toward corporate responsibility with regard to the environment. From your vantage, is most of what we're seeing really green, or just greenwashing? There's surprisingly very little greenwashing out there in the sense of polluting companies being outright misleading. In fact, companies are walking much more than they're talking. Some companies are afraid that talking about what they're doing right will unwittingly illuminate what they're not yet doing right.

Case in point? A few years ago, [one international clothing brand] quietly started sourcing 2 percent of its annual cotton buy organically. They didn't announce the initiative, issue press releases, or identify organic material on product labels. They thought doing so [would risk] customers asking about the other 98 percent being bad for people and the planet. I know of dozens of cases like this.

What sectors have surprised you most in their shifts to greener operations? There are pockets of eco-friendliness in various sectors, often brought about by one leadership company. Interface played a catalytic role in transforming the carpet industry. It didn't take long for its biggest competitors to catch up. Herman Miller has played a similar role in the furniture industry. These are surprising in that neither industry was seen as a big polluter or had been targeted by

Q & A (continued)

activists, shareholders, or others to improve their environmental performance.

Which will have a greater effect on the marketplace, a massive rise in purely green companies or the greening of existing corporate giants? There's no question that, in the short term, the latter represents the biggest opportunity, both for the economy and the planet. There is a tremendous amount of low-hanging fruit in traditional manufacturing operations that can be significantly improved, and profitably. Pure-play green companies play an important leadership role, but not many of them are manufacturers of aluminum, steel, automobiles, cement, and other heavy industry.

If you could pass any law to regulate business, what would it be? No question, it would be to put a price on carbon and other greenhouse gases. Doing so would lead companies of all sizes and sectors to dramatically rethink their operations. Regulating carbon—either through taxes or a cap-and-trade system—would be the number one thing that would drive the economy to be more efficient and more environmentally responsible.

What are the simplest and most effective measures people can take when it comes to greening their work lives? People who drive to work likely have a greater environmental impact getting to and from work each day than anything they do on the job, so finding alternative means of commuting—or, better, telecommuting—is the main thing. Beyond that, they can look for waste and ineffi-

Q & A (continued)

ciency. See what goes into the trash can and Dumpster each day.

Any tips for readers trying to land greener jobs? If you want to be a change agent, look beyond a company's environmental department. First of all, you won't find many jobs there. Most are ghettos of engineers and bureaucrats with little influence. If you want to be a green professional, develop skills, knowledge, and expertise beyond the environment. Learn about marketing, corporate strategy, chemistry, or whatever. Then apply your environmental passions to that arena. If you approach it that way, the opportunities are endless.

Joel Makower is cofounder and executive editor of GreenBiz.com.

Save the Planet in Thirty Minutes or Less

reduces CO_2
improves health
saves money
saves time

- Bring your lunch to work in reusable containers, which cuts down on packaging waste. Keep reusable flatware and cups at the office. $
- Think twice before you print out e-mails and other documents.
- Set up a paper recycling station beside your desk.
- Switch to using recycled paper in your home office.
- Engage the power-saving devices on your computer and other office appliances.

So You Want to Do More

reduces CO₂

improves health

saves money

saves time

- Start reducing junk mail. Whether it's at home or work, disengage from at least one mailing list this week. It's as simple as calling the 800 number on the back of a catalog. For a reasonable fee, services such as GreenDimes and 41pounds will help keep your name off mailing lists. New American Dream and other websites have (ironically) a form letter you can print out to request to be taken off lists.

- Start an office recycling program. Earth 911's website provides a checklist for getting started. GreenBiz.com has tips on products, practices, and more.

- Look for and encourage the purchase of office products and packaging made from recycled and recyclable materials.

- Telecommute at least one day per week, if possible.

- You've done it at home, now do it at work: Get your workplace to switch to safer cleaners, which will improve indoor air quality.

- Always recycle e-waste properly. Earth 911's website lists locations in your community; some big-box stores, including Staples and Best Buy, will take back electronic devices and parts—including ink cartridges, which should also be recycled. Encourage your workplace to do the same.

- Looking for a greener job? TreeHugger.com's Job Board is the place to start pounding the virtual pavement. Visit PlanetGreen.com to connect with other users who are taking steps to green their workplaces.

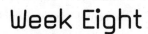

10

Week Eight

Living It Up:
R&R, Volunteering, and Activism

Your mission:
Get out, get busy, speak up.

From the mountains of Colorado to the discotheques of Ibiza, TreeHuggers like to play. Unfortunately, entertaining ourselves isn't always easy on the Earth. Even enjoying the great outdoors can sometimes mean hurting it. Flying to visit distant places contributes carbon dioxide to the ever-growing greenhouse gas overload; driving does the same. But enjoying foreign locales, distant cultures, and natural places is a lot more likely to invigorate and inspire stewardship than surfing the Internet, especially if you're trying to pass along an appreciation of nature and special places to a younger generation. So go ahead and get out there. But remember that how you arrive is likely to make the biggest environmental impact of your trip. Remember that you don't always have to go far to feel far away. Think global, but play local as often as you can, and save some of the emissions associated with long-distance trips.

STAYING GREEN ON THE GO

Eco-resorts, spas, and hotels all around the globe are beginning to offer guests exotic but responsible vacations. When you get to your destination of choice, be it near or far, don't send your green habits on a vacation. If possible, ditch renting a car and use public transit or your own two legs instead. If you must hire a vehicle, look into a hybrid or EV. Car rental services such as Enterprise, Budget, and Avis are expanding their alternative selections. Independent operations are popping up, too. Bio-Beetle, for example, has a biofuel-powered rental fleet in Los Angeles and Maui. Zipcar allows members to rent vehicles in any city in which it operates; if you belong, consider renting as the need arises, instead of for an entire trip.

> # 55 million:
> Number of Americans who prefer travel experiences that protect the ecological and cultural environment.

Getaway Plan

With gaggles of tourists crisscrossing the planet every year, recreational travel comprises a serious chunk of our global carbon footprint. If you're one of the 43 million U.S. travelers concerned about their impact,[1] seeking out responsible resorts is an important part of your travel itinerary. Such properties help protect and conserve the natural and cultural environment and support their local communities. From luxurious spas to off-the-grid getaways, the eco-resorts listed here allow you—and your conscience—to rest easy.

Daintree Eco Lodge and Spa, Queensland, Australia. Set in thirty acres of a World Heritage National Park—the world's oldest living rain forest—this luxury resort puts up visitors in secluded treetop villas. The lodge minimizes its impact by solar-heating its pool and uses native ingredients grown in the region in its restau-

Getaway Plan (continued)

rant. The Rainforest Villas are made from local timber, while the spa integrates Aboriginal techniques and treatments, as well as products that are 100 percent organic.

EcoCamp, Patagonia, Chile. Set in Torres del Paine National Park, EcoCamp was designed to uphold the nomadic spirit of the ancient inhabitants who once lived off the land here. This adventure destination obtains 100 percent of its energy from renewable resources and composts all organic waste. Expert guides educate visitors about local heritage while trekking through the wilderness.

Maho Bay Camps, St. John, U.S. Virgin Islands. This relaxing and secluded resort located in Virgin Islands National Park features hand-constructed, elevated walkways that ensure visitors can enjoy the property while leaving it virtually undisturbed. The Trash to Treasure program uses waste from the property to supply its arts and crafts programs. A "help yourself" center provides a spot for visitors going or coming to leave or take books, sunscreen, and other staples.

Aspen Mountain and Sky Hotel, Aspen, Colorado. A leader in the snow-sports industry, Aspen Mountain was the first resort to implement a green-building policy and to offset 100 percent of its electricity. It also touts the largest photovoltaic system in the industry, trail-grooming vehicles that run on biodiesel, and a corporate matching program that has assisted employees in donating nearly $1 million for local environmental causes. Just as committed, the nearby Sky Hotel encourages guests to recycle with in-room bins and offers complimentary organic, fair-trade coffee and organic snacks and beverages.

Sadie Cove Wilderness Lodge, Homer, Alaska. With accommodations for up to twelve guests, this intimate, off-the-grid lodge is located in pristine Kachemak Bay State Park, home to a diversity of plants and wildlife, including bald eagles, orcas, otters, mountain goats, and much more. The lodge is constructed from driftwood, and a new tree is planted for each guest who arrives.

Wildebees Ecolodge, Hluhluwe, KwaZulu-Natal, South Africa. Situated in the pristine Zulu Kingdom of South Africa's Elephant Coast, Wildebees offers a diversity of activities such as horseback

Getaway Plan (continued)

riding, hiking, and whale- and turtle-watching. To save energy, lights on the property automatically switch off when not in use and fireplaces replace heating systems; rainwater is also filtered and used for drinking. A percentage of profits is designated to projects such as the Odakaneni Community Ark, a program that provides food, clothes, toys, and education to local HIV/AIDS orphans.

Other services and tours can help you find ecologically minded destinations, lodging, and activities. Sustainable Travel International, for example, is a nonprofit service that helps travelers support environmental conservation, cultural heritage, awareness, and economic development in the places they visit. The group also helps vendors and service providers become more conscientious—and more valuable to tourists—through its Sustainable Tourism Eco-certification Program.

You can use them to book sustainable and responsible travel opportunities at eco-lodges everywhere from Alaska to Zambia; buy local, fair-trade, and hand-crafted products; invest in carbon offsets to counterbalance your travel; or just identify and research responsible tourism businesses. The International Ecotourism Society, a similar agency, connects travelers with eco-friendly tours, lodging, and activities that consider the well-being of local people.

Other companies such as Manaca create more upscale, customized trips, choosing in-country operators, guides, and eco-lodges according to their Eco-Assessment tools and your requests. Committed to protecting the environment, these operators use alternative power and eco-friendly transport, recycle, grow and serve local food, support conservation efforts, and give back to local communities. Use Manaca's site to book your entire trip, or just to browse for lodging that meets your needs.

Tips for Greener Travel

- When making travel arrangements, choose overland wherever possible, preferably using mass transit. A bus or train trumps a car when it comes to emissions. All three save massive amounts of CO_2 compared to flying. Bonus: Land travel allows you to see where you're going.

- Before you leave home, adequately adjust your home's heating or cooling system and turn down the thermostat on your hot-water heater. Unplug all appropriate electronics and stop any newspaper deliveries if you haven't already switched to digital subscriptions. You may be able to donate your papers to a school.

- At hotels, let housekeeping know that it's not necessary to change your towels and sheets every day. Bring your own toiletries. Those small free bottles waste packaging (though some hotels reuse them). Use reusable containers for toiletries instead of buying travel-size items.

- Turn off the air-conditioning or heat, lights, and TV when you leave the room.

- At your destination, whenever possible, walk and use public transportation or the hotel van instead of using a car.

- Take free brochures and maps only as necessary.

- Travel with a reusable water bottle and coffee mug.

- Don't buy products you suspect are made from parts of endangered species, such as tortoise shell, ivory, skins, and feathers. Seashells and other marine life are often also questionable.

- Don't collect natural souvenirs from wild areas, which could disrupt ecosystems.

- Check out of your hotel room via in-room electronic programs, if available, which reduces usage of paper.

Cameras: Film versus Digital

So you've decided to go on an African safari—one where you'll shoot photographs, not animals. Good choice. You've offset your flight, researched eco-friendly lodging, and borrowed rather than bought binoculars for the trip. But now you're probably wondering, "What about my camera?" At first thought, a digital device might appear to be the obvious environmental choice. This would save film from being used and spare the chemicals to develop it. Plus, you can delete or choose not to print any unwanted images. It's a no-brainer, right?

Not quite. A life-cycle assessment study conducted by the Georgia Institute of Technology found that a lifetime of charging up electronic cameras and manipulating photos on computers adds up. The electricity consumed by digital cameras washes out their environmental benefits when compared to conventional ones.[2] Plus, with technology changing so quickly, consumers are replacing electronic gadgets frequently, increasing manufacturing demands and global transportation.

There's no clear winner in this debate, but what you do with your equipment and photos can positively or negatively influence their environmental impact. A couple of tips: Digital camera owners should make their cameras last. Those sticking with traditional tech should buy film rolls with thirty-six exposures, which reduces packaging waste. To further protect the wildlife whose image you wish to capture, steer clear of disposable cameras, which are disparagingly built for one-time use.

THE SPORTING LIFE

Whether trekking the Himalayas or playing tennis on Saturdays, leisure time and recreation are essential human needs. But the gear we use to have our fun can directly and indirectly affect the environment we're enjoying—whether it's a suburban setting or a fragile alpine ecosystem.

The greenest outdoor pursuits are those undertaken without the aid of fuel-guzzling engines. Choosing the crunch of fresh powder under your snowshoes rather than the screech and grind of a snow-mobile is more serene and less polluting. The same thing goes for the dip-and-drip of a paddle over the whirring of personal water-craft, the flapping of a sail over the whine of twin 250s, and dodging twigs and roots on trails over the gym's gerbil-like treadmill.

Various sporting sectors are already bracing for the effects of global warming and other environmental consequences. Some, like the snow-sports industry, argue they're already feeling those effects. Logically, global warming is poised to affect mountain re-sorts early on. Unsurprisingly, then, the industry has been one of the first to respond to the problem. Of course, they've also been one of the culprits. But as sources of fresh water and enormous ge-netic diversity, mountains aren't essential just to skiers; they're crucial to life on earth.

The most obvious consequences of global warming—less snowfall, shorter ski seasons, and higher resort prices—are pretty big and tangible bummers. Some areas that have recently suffered these effects believe that climate change is already here. That's why resorts such as Aspen in Colorado are taking initiatives to support wind energy and convert snowcats to biodiesel. Others, such as the Falls Creek resort in Australia, are buying renewable energy credits, recycling water to make snow, and doing away with plastic bags. Efforts such as these don't give the ski industry a free pass, but they can help raise awareness. Of course, skiers and riders who really want to reduce their footprint can skip lift lines alto-gether and hit the backcountry.

Secondhand Gear

Depending on your objectives, secondhand gear can be a terrific way to lighten your load on the planet. Just need a stick for pond hockey? Trying out squash for the first time? You probably don't need top-of-the-line equipment just yet. Like all sorts of other things pre-loved, many sports articles find a second life through Freecycle, eBay, and Craigslist. Play It Again Sports shops provide aspiring athletes across the nation with the pre-owned equipment they need. End-of-the-season rental shop sales are another good source for slightly used stock.

Before you plunk down the cash to own something, consider how often you'll really use it. It might prove more economically and environmentally sound to rent. Think of it this way: You'll always be skiing on the latest sticks and you won't have to store that sea kayak you use only once a year. Chances are good that no matter what it is, you can rent it.

Speaking of the kayak gathering dust in your garage, why not close the loop and sell or donate yours to someone who really will use it? Organizations such as SWAG (Sharing Warmth Around the Globe) distribute retired ski clothing to people in need. Porters' Progress sends boots, sleeping bags, and outerwear to the Nepalese. The Mountain Fund passes along packs to impoverished people. Recognize that your trash is someone else's treasure. Scout troops, Boys and Girls Clubs of America, and other local groups are likely to accept and redistribute things closer to home.

Extending the life of the gear you decide to hang on to is useful, too. Patch that surfboard ding; seal that tent tear; re-waterproof your rain jacket. (Nikwax makes a water-based water-repellent that is easy to use and claims to be environmentally friendly.) If the job is too big for you to do yourself, use a repair service. The point is that making your gear last as long as possible can go a long way toward helping out both you and the planet. So rinse those climbing ropes, restring that racquet, and spring for that camp stove cleaning kit.

When Only New Will Do

There will always be times when nothing but brand-new helmets, harnesses, or lines will do. Obviously it's important to select gear for its ability to perform—in some cases, your safety or your life may depend on it. Increasingly, new products are available that offer identical function—and flair—with fewer environmental flaws. Bamboo snowboards, hemp surfboards, recycled plastic underwear and fleece, recycled rubber-soled shoes, and organic cotton pants are all examples of sports equipment and apparel that present a good way to get a little closer to nature without taking its life in your hands.

National retailer REI has been progressively introducing greener products, programs, and even ecologically designed, LEED-certified stores. Recently, the brand rolled out its Eco-Sensitive label, whose products are made from sustainable, organic, or recycled materials. To help climate-control yourself and the planet, technical and casual apparel maker Nau uses recycled polyester, cornstarch, and organic cotton to create TreeHugger-approved, functional looks. Ibex and Smartwool pass over petroleum products altogether to offer all-natural wool-base layers; Teko makes high-performance socks from sustainable wool, recycled polyester, Ingeo (a cornstarch polymer), and organic cotton.

Gear and accessories manufacturers are following suit. Arbor, Indigo, and Venture Snowboards craft boards and skis from bamboo and other sustainable materials. Hess Surfboards foregoes industry-standard foam blocks made from fossil fuel by-products and instead opts for recyclable foam and cork, and frames built from reclaimed and sustainable timber. Hillbilly Wax-Works' Enviro-wax for skis and snowboards is biodegradable and free of toxins; Cradle to Cradle–certified Wet Women Surf Wax (it works for guys, too) is also eco-safe. Want more? Pedro's makes environmentally friendly bike care products, such as Green Fizz foaming bike wash. Patagonia makes a wet suit from nonpetroleum-based neoprene, recycled polyester, and wool.

Across the sports industry we are seeing skateboards, climbing gear, running shoes, tents, kayaks, solar gadget chargers, and other products reach innovative new eco-friendly highs. If you cannot find an ecological equivalent to something you must have, look for brands that are at least involved in environmental stewardship. Many companies belong to programs such as One Percent for the Planet, an alliance of hundreds of businesses that contribute a portion of their sales to grass-roots environmental groups.

CASE STUDY: Patagonia

Any discussion of sports and the environment would not be complete without devoting space to Patagonia, which has led the pack producing green gear for a wide swath of the market, from surfing to mountaineering. But although the private company is renowned for a visionary business strategy and environmental commitment, Patagonia founder and owner Yvon Chouinard didn't make a conscious decision to become a green business guru.

When Chouinard began big-wall climbing in the 1950s in places such as Yosemite, the only pitons (the spikes climbers drive into the rock) available were designed for one-time use. So Chouinard decided to make his own reusable hardware. His friends were impressed, and a business enterprise was born.

Fast forward a few decades, and Patagonia had become one of the most popular names in outdoor equipment, respected for its innovative garments, such as moisture-wicking Capilene base layers made from recycled polyester and Synchilla fleece made from recycled plastic bottles. Today, through its Common Threads recycling program, Patagonia allows customers to return its T-shirts

(Sources: TerraPass; "Solutions: A Hiker's Guide to Fighting Climate Change,"

CASE STUDY: Patagonia (continued)

and base layers and any brand of fleece to the company to be refashioned into new fibers and garments.

Turning the lens on its own operations and procedures has been paramount. After learning that its cotton garments had a higher environmental impact than its oil-based polyester and nylon ones, Patagonia switched to 100 percent organic cotton in 1996, and never looked back. At the time, there was so little of it commercially available that they had to source it directly from farmers and then persuade ginners and spinners to adjust their equipment.

In the 1970s, Patagonia established a habit of making regular financial contributions to small environmental groups working to save or restore habitats, rather than big national or international nonprofits. That commitment has evolved into donating 10 percent of profits or 1 percent of sales, whichever is higher, every year to support such organizations. The principle led Chouinard, in 2001, to co-found One Percent for the Planet, a global program that provides a platform for responsible business philanthropy in support of environmental causes. To date, members have contributed more than $21 million.

The list of details that contribute to Patagonia's green reputation goes on—catalogs printed on recycled, FSC-certified paper; a LEED-certified distribution center; the purchase of renewable electricity for stores and offices. Chouinard has indeed followed his mission to use his business "to inspire and implement solutions to the environmental crisis." His long-standing and deep pledge to corporate environmental behavior has left a mark on the sporting-goods industry and consumers. Fortunately, it's a mark that treads lightly on the planet.

HOLLYWOOD LENDS A HAND

Even if you're not addicted to magazines such as *In Touch* and *Us Weekly,* you probably know that a lot of red-carpet walkers and rock royalty are talking up the importance of taking care of the

by Berne Broudy, *Backpacker,* September 2007)

planet. True, the Hollywood lifestyle and blockbuster films can leave a big footprint. But trendsetting is something stars do very well, and the attention some have lavished on living green is trickling down and affecting the behavior of the general public. From tour buses and generators that run on biodiesel to production companies that offset the carbon costs of their films to solar-

Green Carpet Treatment

Match the celebrity with the environmental cause:

1. Brad Pitt

a. Backyard Habitat

b. WildAid

2. Julia Louis-Dreyfus

c. Global Green's Sustainable Design Competition for New Orleans Neighborhoods

3. Willie Nelson

d. The United Nations Water for Life program

4. Yao Ming

e. Earth Biofuels

5. Martha Stewart

f. The Trust for Public Land

6. Jay-Z

Answers on page 202.

powered recording studios, Hollywood is making some noise. When celebs such as Cameron Diaz, Bono, and George Clooney open their mouths or arrive at awards show in Priuses, people pay attention.

Taking an anti-*Cribs* approach, HGTV's *Living with Ed* focuses on how Ed Begley passed up the typical Hollywood "good" life for a better, greener one: He lives large in a small house, drives an electric car, and runs his toaster off power generated while he rides his stationary bike. Even MTV's *The Real World* has gone green. And who would have thought that a PowerPoint presentation about climate change known as *An Inconvenient Truth* would win two Academy Awards?

The trend extends beyond Tinseltown. Tennessee's Bonnaroo Music and Arts Festival uses biodegradable concession service items and cleaner power options and sells organic and hemp T-shirts. Festival garbage is recycled into park benches to be used at future events. At Bumbershoot, Seattle's music and arts festival, organizers go beyond carbon-neutral to offer carpoolers discounted parking, provide bike parking, use signage made from biodegradable materials, print concert literature on FSC-certified paper, use green power, and more.

Entertaining Yourself

You might not be a star who can show off your eco-threads on-stage, make a statement on the cover of *Vanity Fair*, get cameras to capture you washing your roof's PV panels, or use your mike to reach a few million fans. But when it comes to entertainment, you can make a difference by using smart technology that reduces packaging, shipping, and waste. By downloading music and videos instead of buying CDs and DVDs, for example, you can help save resources. Renting or sharing DVDs also cuts down on the amount of physical stuff we consume—and we can guarantee that watching your neighbor's copy of *Caddyshack* will be just as amusing as if you owned it yourself.

Attending concerts, festivals, and movies that act out of concern for the environment is another way to make a difference. Tell them why you're there. It can be tough to get a message directly to Leo DiCaprio, sure, but you can let local entertainment venues, radio stations, newspapers, and other moviegoers and concertgoers know what you think. Entertainers have a powerful knack for getting heard and spreading the word; let them know what you want them to talk about.

Green Media: Books, Movies, and Music

BOOKS: Essentials for Your Library

Natural Capitalism: Creating the Next Industrial Revolution. Paul Hawken, Amory Lovins, and L. Hunter Lovins (Back Bay Books/Little, Brown and Company, 1999/2000). A groundbreaking look at how the successful businesses of the future will profit from their own environmental responsibility.

Cradle to Cradle: Remaking the Way We Make Things. William McDonough and Michael Braungart (North Point Press, 2002). An architect and a chemist unite to explore how the manufacturing industry can emulate nature, where the concepts of waste and pollution are nonexistent.

Silent Spring. Rachel Carson (Marine Books, 1962/2002). This landmark book launched a revolution by sounding the alarm about pesticide use and environmental devastation.

The Weather Makers: How Man Is Changing the Climate and What It Means for Life on Earth. Tim Flannery (Atlantic Monthly Press, 2006). Flannery presents compelling scientific evidence that climate change is real and happening.

Green Media (continued)

Plan B 2.0: Rescuing a Planet Under Stress and a Civilization in Trouble. Lester R. Brown (W. W. Norton, 2006). Brown presents his blueprint for saving modern civilization through a global cooperative effort.

The Creation: An Appeal to Save Life on Earth. E. O. Wilson (W. W. Norton, 2006/2007). In a series of letters to a Southern Baptist pastor, Pulitzer Prize–winning entomologist Wilson argues that religion and science must unite to preserve all of creation.

Your Money or Your Life: Transforming Your Relationship with Money and Achieving Financial Independence. Joe Dominguez and Vicki Robin (Penguin, 1999). This nine-step program for personal financial transformation is a highly effective tool for shrinking your footprint, too.

The Lorax. Dr. Suess (Random House Books for Young Readers, 1971). Long before "global warming" was a household term, the Lorax was priming kindergartners to understand the dangers of mindless progress.

The End of Nature. Bill McKibben (Random House Trade Paperbacks, 1990/2006). McKibben's classic explores the environmental crisis of our era and the need for a fundamental shift in the way humans relate to nature.

Stuff: The Secret Lives of Everyday Things. John C. Ryan and Alan Thein Durning (Northwest Environment, 1997). A behind-the-scenes look at the energy-intensive and environmentally unsound backgrounds of seemingly innocent everyday objects.

WorldChanging. Alex Steffen, ed. (Abrams, 2006). An encyclopedia of bleeding-edge tools, technologies, and ideas.

MOVIES: Ones to Watch

Silent Running (1972)

Chinatown (1974)

Koyaanisqatsi (1982)

Erin Brockovich (2000)

The Day After Tomorrow (2004)

> ## Green Media (continued)
>
> *An Inconvenient Truth* (2006)
> *Who Killed the Electric Car?* (2006)
> *Arctic Tale* (2007)
>
> ### SONGS: Music to Listen Up By
> "Big Yellow Taxi," Joni Mitchell
> "Mercy Mercy Me," Marvin Gaye
> "(Nothing But) Flowers," Talking Heads
> "My City Was Gone," The Pretenders
> "The 3 R's (Reduce, Reuse, Recycle)," Jack Johnson
> "Rocky Mountain High," John Denver
> "Throwing Stones," The Grateful Dead
> "If a Tree Falls," Bruce Cockburn
> "With My Own Two Hands," Ben Harper
> "After the Garden Is Gone," Neil Young

BECOMING ACTIVE

Eco-activism groups such as Earth Liberation Front and Earth First! had their place in raising awareness about serious environmental issues in the latter decades of the last century. But let's face it, tree-spiking and roadblocks are *so* 1987. The counterculture methodologies and sometimes-destructive behavior of radical eco-warriors were often counterproductive to their goals, not to mention their mainstream popularity.

Today's greens have embraced new tools for spreading the word. We don't have to sabotage buildings or torch a fleet of Hummers. (That kind of destruction only wastes valuable resources and materials.) Technology has made virtual tree-hugging an extremely effective tool for connecting people, building support, getting voices heard, effecting change, *and* saving fuel on travel to and from demonstrations. Contemporary activists have shed their image of desperation for one of hopefulness; dangerous stunts have been supplanted with playful antics. To wit, guerrilla gardeners—rebels who convert neglected wisps of land into flower beds—have reared their gloves from Toronto to London. Eco-artists have turned metered parking spots into temporary parks, complete with sod and benches.

The Internet has provided organizations such as the Natural Resources Defense Council a central location from which not only to spread its message but also to allow concerned citizens to get their collective voices heard. Tackling topics such as global warming, clean air and water, and endangered ecosystems, the non-profit has fought to protect human health and the natural world since 1971.

Letter-writing campaigns have long been proven means for effecting change, and groups such as the NRDC make it easy to participate. They write the form letter; you sign off by attaching your e-mail address. With the help of supporters, during the past few years the organization has helped safeguard polar bears, prevented oil drilling in Alaska, and initiated cleaner vehicles programs. If that's the kind of stuff you care about, join their website's Activist Network to receive e-mail updates about pertinent issues. The site also provides

Eco-myth:
Activism means marching in the streets, chaining yourself to a tree, or getting politically involved.

Fact:
Activism is anything you do to make the world a little greener. That could mean telling friends about energy-efficient lighting, encouraging your child's school to improve indoor air quality, or shopping at a farmers market.

practical tips for writing, phoning, and e-mailing your legislative representatives directly.

An organization with a similar mission, the Union of Concerned Scientists is a watchdog group launched in 1969 as a joint initiative between students and professors at the Massachusetts Institute of Technology. Their efforts focus on topics such as food safety, nuclear policy, clean air and vehicles, and global warming. Through independent scientific research and citizen action, UCS works to develop realistic solutions and changes in government policy, corporate practices, and consumer choices. Like the NRDC, the UCS makes it easy for you to stay abreast of topics and make your voice heard. Its online Action Center provides sound advice for things such as contacting your legislators and raising issues at public meetings.

Getting involved at the local level can be an accessible and gratifying way to connect with your community and bring about grass-roots change. Think global, act local. It's not just a cliché; it's a manifesto. But if even nonpartisan political action makes your palms sweat, you might consider volunteering with or joining a group such as Surfrider Foundation, a national agency with local chapters that help people connect with nature. Thousands of independent organizations with similar goals exist across the country. But you don't have to be officially affiliated with any entity to help out. You can join a beach cleanup, tell a neighbor about the green cleaning products you're using, or just pick up litter when you're walking the dog. It's that simple to spread the word and be the change. No evangelism necessary.

> This is the first age that's ever paid much attention to the future, which is a little ironic since we may not have one.
> —Arthur C. Clarke

GETTING TO GRASS-ROOTS CHANGE

There's no law that says you have to commune with nature or write to your senator just because you care about the environment. If your idea of getting outdoors is walking to the bus stop, trail maintenance days and mountaintop rallies probably aren't for you. Stop Global Warming's Virtual March on Washington allows you to demonstrate metaphorically, without having to leave the comfort of your office. Launched by Hollywood producer Laurie David, this nonpartisan site is working to bring citizens together to declare that global warming is an urgent reality that requires immediate attention.

Close to a million registrants—Arnold Schwarzenegger and John McCain among them—have made a virtual commitment to enact change as individuals, and their collective numbers suggest they really mean it. Laurie David certainly does. As the brains behind *An Inconvenient Truth*, she knows a thing or two about reaching the masses, and she's used her connections and entertainment know-how to fuel events such as the Stop Global Warming College Tour with Sheryl Crow.

Regardless of how you do it, participation—no matter how small—is what TreeHugging is really all about. Identify what you're willing to do. Whether it's donating money to a nonprofit, taking a volunteering vacation, attending a local fund-raiser, or taking part in TreeHugger's online forums, there are myriad ways to contribute.

None of us is in this alone. Becoming greener is about discovering the interconnectedness of all things. The world's a small, intimate place. Shaping its communities is up to each of us. Will you spread the word? Plant a tree? Switch to organic applesauce? For better or worse, no action is without an impact. At its most basic, being green means having an awareness that the environment is all around us. It depends on us—and we, on it.

Q & A

Laurie David, Founder of Stop Global Warming, on the Power of Hollywood and Why Everyone Is an Environmentalist

Before you became a climate change advocate, you worked behind the scenes in Hollywood. What made you feel you needed to get on the front lines? The main event in my life was becoming a mother and suddenly having responsibility for these creatures. That coincided with the explosion of SUVs; everybody I knew drove them, and it was very obvious that they were big polluters. The more I learned, the more upset I got. Once I started to connect the dots with climate change, there was no going back.

What drives your activism? I really view this as a life-and-death issue. We need massive change, and it's going to come from individuals pushing up against their families, families against business, and business against government.

What can one person do? That's the exciting thing: There aren't many issues where there's actually something you can do about it, but since we're all guilty of contributing to the problem of global warming, we can all be part of the solution. What can one person do? Try changing a lightbulb. Multiply that by millions, and you have a serious force. All the small actions add up.

Why an online demonstration? In the old days, we'd march out on the streets and it turned into a thirty-second clip on the evening news. But this problem is so huge and urgent that we need to march every single day. How do you tackle modern problems? With modern technology.

Q & A (continued)

What role does Hollywood have to play in counteracting global warming? Two things that celebrities do well are to bring attention to a cause and help set trends. The hybrid car is a perfect example. These weren't the hippest, coolest cars, but because celebrities started driving them, other people became interested. Movies geared toward kids, like *Ice Age* and *Arctic Tale*, are educational; kids go home and put pressure on their parents. Hollywood also made *An Inconvenient Truth* and *The Day After Tomorrow*. When the latter came out, every morning show in the country did a story on global warming. Hollywood deserves a lot of credit.

Speaking of *An Inconvenient Truth*, which you helped produce, did you ever think it would have such a huge impact? I never, ever thought of the film in relation to the Academy Award, but I did know in my gut that this film would make a difference. So I was not surprised when it did well. The first time I saw a four-minute version of Al Gore's presentation, my jaw dropped. I knew that it was the single best way to explain this information and that if people saw it, they'd respond.

How do you counterbalance your own carbon footprint? Nobody can survive in this world without being a carbon emitter. The goal is reduction. It's not about sacrifice; it's about change. I drive a hybrid car, buy offsets, support sustainable industry through my stock purchases, reduce, reuse, recycle, carry a metal water bottle, use solar panels—the list goes [on]. I'm not perfect. I do the best I can, and then I try to do better.

Q & A (continued)

You once said that everyone is an environmentalist, but some people just don't know it yet. What did you mean?
Show me the person who doesn't enjoy a beautiful spring day, riding a bike, swimming in the ocean, or skiing. Who doesn't love fall leaves changing? Who doesn't want to drink clean water? If you like these things, you're an environmentalist. There's nothing political about it, either. There are solutions that are political, but the core issue is not. Environmentalism isn't a special group that you join; it's in your heart and soul.

Laurie David is co-author of the kids' book *The Down-to-Earth Guide to Global Warming* (Scholastic, September 2007).

Save the Planet in Thirty Minutes or Less

reduces CO$_2$

improves health

saves money

saves time

- When traveling, choose the least polluting transit option. Mass transit such as the bus, train, or ferry are better options than driving alone in your car. If you must fly, consider buying carbon offsets to negate the unavoidable greenhouse gas emissions related to your leisure travel. Some websites such as Expedia and Travelocity offer to tack on carbon credits at the point of sale, or you can use an offset provider like TerraPass.

- Pledge to make at least one trip next year be one that's closer to home—even if it's just a weekend jaunt.

- Start making it a habit to download music instead of purchasing CDs. You may also be saving on transportation costs associated with driving to the store.

- Take a few minutes to enjoy the great outdoors. It doesn't matter if you live in Manhattan or rural Kansas or whether you're walking to the subway, hiking through a forest, or swinging a racquet on a tennis court. Stop. Look around. Notice. Smell. The natural world, even if you're in an urban jungle, is all around you. ♡

- Are you ready to get more involved in your community? If so, identify your interests, evaluate your potential time commitment, and define what you can give (money, time, effort, in-kind donations). Go for it. 🗲 ♡

- Join the Virtual March on Washington at StopGlobalWarming.org. 🗲 ☉

So You Want to Do More

🗲 reduces CO_2

♡ improves health

$ saves money

☉ saves time

- Fly direct whenever possible. Doing so saves fuel, as more is needed for takeoffs and landings. 🗲

- Volunteering vacations are a terrific—and typically low-cost—way to get up close and personal with remote areas. From saving turtles in Ecuador to counting birds in Bhutan to building houses in Alaska—these hands-on experiences can be a lot more fruitful than sipping piña coladas on the beach. ♡ $

- Were you inspired by the Academy Award–winning film *An Inconvenient Truth*? Consider becoming a Climate Project ambassador and learn how to tell it like Al. 🗲

- Spread the word. If people are curious, tell them how and why you bike to work, compost, or buy organic. Take it a step further and organize community discussion groups or workshops. 🗲

- Clean the sports equipment out of your garage or attic. Donate it to a place that will use it. 🗲 $

- Bring your own food and water with you when you travel. You'll cut down on waste and be eating healthier snacks.

- Celeb-obsessed? Got wanderlust? Stay informed about what the rich and famous are doing to save the planet and where the international hotspots are at TreeHugger.com. Connect with other greens to gossip or get involved at PlanetGreen.com.

resources and further reading

2. PREP WORK—GETTING STARTED

Global Footprint Network (www.footprintnetwork.org)
Environmental Defense's Fight Global Warming
    (www.fightglobalwarming.com).
An Inconvenient Truth. Al Gore (Rodale, 2006).

3. WEEK ONE—THINKING LIKE A TREEHUGGER

Biomimicry: Inspiration and Innovation in Nature. Janine Benyus
    (HarperPerennial, 2002).
Cradle to Cradle: Remaking the Way We Make Things. William McDonough and
    Michael Braungart (Northpoint Press, 2002).
Earth 911 (www.earth911.org).
The Internet Recycling Guide (www.obviously.com/recycle).
Recycle: The Essential Guide. Duncan McCorquodale, Cigalle Hanaor, and Lucy
    Siegle (Black Dog Publishing, 2006).

4. WEEK TWO—EATING YOUR WAY GREEN

100 Mile Diet (www.100milediet.org).
American Grassfed Association (www.americangrassfed.org).
Deep Economy: The Wealth of Communities and the Durable Future. Bill McKibben
    (Times Books, 2007).
The Eat Well Guide (www.eatwellguide.org).
Local Harvest (www.localharvest.org).
Monterey Bay Aquarium's Seafood Watch (www.seafoodwatch.org).
The Organic Trade Association (www.ota.com).
Slow Food Movement (www.slowfoodusa.org).
The Union of Concerned Scientists Action Center (www.ucsaction.org/
    ucsaction).

5. WEEK THREE—GREENING UP YOUR ACT

Center for Children's Health and the Environment (http://childenvironment
    .org).
The Complete Organic Pregnancy. Deirdre Dolan and Alexandra Zissu (Collins,
    2006).

Environmental Working Group's Human Toxome Project (www
.bodyburden.org).
Greenguard Environmental Institute (www.greenguard.org).
Green Seal (www.greenseal.org).
Habitat for Humanity (www.habitat.org).
How to Grow Fresh Air: 50 Houseplants That Purify Your Home or Office.
B. C. Wolverton (Penguin, 1997).
Material Safety Data Sheets (www.msdssearch.com).
National Institute of Health Household Products Database
(http://hpd.nlm.nih.gov).
Naturally Clean: The Seventh Generation Guide to Safe & Healthy, Non-Toxic Clean-
ing. Jeffrey Hollender, Geoff Davis, with Meika Hollender, and Reed
Doyle (New Society Publishers, 2006).
Pittsburgh Toy Lending Library (www.pghtoys.com).
Seventh Generation (www.seventhgeneration.com).

6. WEEK FOUR—TRAVELING LIGHT

Clean Air Cool Planet (www.cleanair-coolplanet.org).
Earth Policy Institute (www.earth-policy.org).
Gold Standard Foundation (www.cdmgoldstandard.org).
Greasecar (www.greasecar.com).
Green Car Congress (www.greencarcongress.com).
The National Biodiesel Board (www.biodiesel.org).
Plan B 2.0: Rescuing a Planet Under Stress and a Civilization in Trouble. Lester
Brown (W. W. Norton, 2006).
U.S. EPA Green Vehicle Guide (www.epa.gov/emissweb).
Zipcar (www.zipcar.com).

7. WEEK FIVE—THE GREENER HOME

Energy Star (www.energystar.gov).
Green-e (www.green-e.org).
Home Energy Saver (http://hes.lbl.gov).
Rocky Mountain Institute (www.rmi.org).
U.S. Department of Energy: A Consumer's Guide to Energy Efficiency and
Renewable Energy (www.eere.energy.gov/consumer).
U.S. EPA WaterSense (www.epa.gov/watersense).
U.S. Green Building Council (www.usgbc.org).

8. WEEK SIX—DRESSING UP

Awakening Beauty the Dr. Hauschka Way. Susan West Kurz (Clarkson Potter,
2006).
Bag Borrow or Steal (www.bagborroworsteal.com).
Consumer Reports Greener Choices (www.greenerchoices.org).

The Consumers Union Guide to Environmental Labels (www.eco
    -labels.org).
Green Earth Cleaning (www.greenearthcleaning.com).
The Laundress (www.thelaundress.com).
Liquid CO$_2$ cleaners (www.findco2.com).
Organic_Clothing (www.organicclothing.blogs.com).
Skin Deep, Environmental Working Group's Cosmetics Safety Database
    (www.cosmeticsdatabase.com).
Swaporamarama (www.swaporamarama.org).

9. WEEK SEVEN—GETTING TO WORK

Call2Recycle (www.rbrc.org/call2recycle).
Dow Jones Sustainability Indexes (www.sustainability-indexes.com).
The Ecology of Commerce. Paul Hawken (Collins, 1994).
Electronic Product Environmental Assessment Tool (www.epeat.net).
GreenBiz (www.greenbiz.com).
Green Money Journal (www.greenmoneyjournal.com).
The Green Office (www.thegreenoffice.com).
Natural Capitalism: Creating the Next Industrial Revolution. Paul Hawken, Amory
    Lovins, L. Hunter Lovins (Back Bay Books, 2000).
ResponsibleInvesting.org (www.responsibleinvesting.org).
Social Investment Forum (www.socialinvest.org).
TreeHugger Job Board (http://jobs.treehugger.com).
U.S. EPA's Indoor Air Quality (www.epa.gov/iaq).

10. WEEK EIGHT—LIVING IT UP

One Percent for the Planet (www.onepercentfortheplanet.org).
G-ForSE (www.g-forse.com).
Green Drinks (www.greendrinks.org).
An Inconvenient Truth (www.climatecrisis.net).
The International Ecotourism Society (www.ecotourism.org).
The Leave No Trace Center for Outdoor Ethics (www.lnt.org).
Manaca (www.manaca.com).
Natural Resources Defense Council (www.nrdc.org).
Stop Global Warming (www.stopglobalwarming.org).
Sustainable Travel International (www.sustainabletravel.com).
Union of Concerned Scientists (www.ucsusa.org).

notes

2. PREP WORK—GETTING STARTED

1. Intergovernmental Panel on Climate Change, Fourth Assessment Report, Working Group III, "Climate Change 2007: Mitigation of Climate Change, Summary for Policy Makers" (Geneva, 2007), p. 10.

2. Michael Le Page, "Climate Change: A Guide for the Perplexed," *New Scientist* (May 16, 2007), http://environment.newscientist.com/channel/earth/dn11462.

3. Naomi Oreskes, "Beyond The Ivory Tower: The Scientific Consensus on Climate Change," *Science* 306, no. 5702 (December 3, 2004): 1686, http://www.sciencemag.org/cgi/content/summary/306/5702/1686.

3. WEEK ONE—THINKING LIKE A TREEHUGGER

1. U.S. Environmental Protection Agency, "Municipal Solid Waste in the United States: 2005 Facts and Figures," (October 2006), p. 12.

2. Ibid., p. 5.

3. U.S. EPA, *Consumer's Handbook for Reducing Solid Waste* (September 2007), http://www.epa.gov/epaoswer/non-hw/reduce/catbook/debate.htm.

4. U.S. EPA, "Municipal Solid Waste," pp. 7–8.

5. U.S. EPA, "Municipal Solid Waste," p. 9.

6. William McDonough and Michael Braungart, *Cradle to Cradle: Remaking the Way We Make Things,* (New York: Northpoint Press, 2002), pp. 27–28.

7. Ibid., p. 28.

8. Worldwatch Institute, "Plastic Bags," http://www.worldwatch.org/node/1499.

9. United Nations Environment Programme, "The Packaging Nightmare" (2006), http://www.vitalgraphics.net/waste2/.

10. United Nations Environment Programme, "Call for Global Action on E-waste," December 1, 2006, http://www.unep.org/Documents.Multilingual/Default.asp?DocumentID=496&ArticleID=5447&l=en.

11. Matthew D. Sarrel, "Recycling E-Waste," *PC Magazine* (November 29, 2006), http://www.pcmag.com/article2/0,1895,2064151,00.asp.

12. "The Truth About Recycling," *The Economist,* June 7, 2007.

13. U.S. EPA, "Puzzled About Recycling's Value?" p. 1.

14. Ibid., p. 9.

15. U.S. EPA, "Municipal Solid Waste," p. 11.

16. McDonough and Braungart, *Cradle to Cradle*, p. 91.

17. Janine M. Benyus, *Biomimicry*, (New York: Perenniel, 2002 [1997]), pp. 132–35.

4. WEEK TWO—EATING YOUR WAY GREEN

1. Gidon Eshel and Pamela A. Martin, "Diet, Energy, and Global Warming," *Earth Interactions* 10 (May 2005).

2. Janet Larsen, "Dead Zones Increasing in World's Coastal Waters," Earth Policy Institute, June 16, 2004, http://www.earth-policy.org/Updates/Update41.htm.

3. United States Department of Agriculture, "USDA Releases New Farmers Market Resource Guide," March 13, 2006, http://www.usda.gov/wps/portal/!ut/p/_s.7_0_A/7_0_1OB?contentidonly=true&contentid=2006/03/0081.xml.

4. Organic Trade Association, "Organic Agriculture and Production," http://www.ota.com/definition/quickoverview.html.

5. The Container Recycling Institute.

6. Children's Health Environmental Coalition, "Feeding with the Bottle," http://www.checnet.org/HealtheHouse/education/articles-detail.asp?Main_ID=333.

7. Eshel and Martin, "Diet, Energy, and Global Warming."

8. Union of Concerned Scientists, "New USDA Grass-Fed Rules Will Benefit Consumers and the Environment, October 17, 2007, http://www.ucusa.org/news/press_release/rules.html.

9. Ibid.

10. Editors of *E the Environmental Magazine*, *Green Living: The E Magazine Handbook for Living Lightly on the Earth*, (New York: Plume, 2005), p. 3.

11. Union of Concerned Scientists, "Scientists Say Cloning Animals for Food Has Uncertain Benefits, Many Drawbacks," December 28, 2006, http://www.ucsusa.org/news/press_release/scientists-say-cloning.html?wt.rss=rss.

5. WEEK THREE—GREENING UP YOUR ACT

1. Jeffrey Hollender, Geoff Davis, with Meika Hollender and Reed Doyle, *Naturally Clean: The Seventh Generation Guide to Safe & Healthy, Non-Toxic Cleaning* (Gabriola Island, BC: New Society Publishers, 2006), pp. 8, 20; Board on Health Sciences Policy, *Cancer and the Environment: Gene-Environment Interactions* (Washington, D.C.: National Academies Press, 2002), p. 26, http://books.nap.edu/openbook.php?record_id=10464&page=25.

2. Hollender et al., *Naturally Clean*, p. 44.

3. Environmental Working Group, "Body Burden—The Pollution in Newborns" July 14, 2005, http://www.archive.ewg.org/reports/bodyburden2/part3.php.

4. Silent Spring Institute, "Endocrine Disruptors in Indoor Air and Dust in Cape Cod, MA, Homes," (January 2004) http://www.silentspring.org/newweb/research/HES%20Long%20Q&A.pdf.

5. Madeleine Cobbing, "Cleaning Up Our Chemical Homes: Changing the Market to Supply Toxic-Free Products" (Amsterdam: Greenpeace International, February 2007), p. 11; Peter Montague, "Human Breast Milk Is Contaminated," *Rachel's Democracy and Health News,* August 8, 1990, http://www.rachel.org/bulletin/index.cfm?St=4.

6. Cobbing, "Cleaning up Our Chemical Homes," p. 16.

7. U.S. Environmental Protection Agency, "Ozone Generators That are Sold as Air Cleaners" (August 2007), http://www.epa.gov/iaq/pubs/ozonegen.html.

8. "Many Cleaners, Air Fresheners May Pose Health Risks When Used Indoors," *ScienceDaily,* May 24, 2006, http://www.sciencedaily.com/releases/2006/05/060524123900.htm

9. Cobbing, "Cleaning Up Our Chemical Homes," p. 27.

6. WEEK FOUR—TRAVELING LIGHT

1. New American Dream, http://www.newdream.org/make/auto/facts.php.

2. John Davies, "Transportation Greenhouse Gas Emissions 1990–2003: Trends, Uncertainties and Methodological Improvements" (U.S. Environmental Protection Agency Transportation and Climate Division Office of Transportation and Air Quality, May 18, 2006), p. 9.

3. Lester R. Brown, *Plan B 2.0: Rescuing a Planet Under Stress and a Civilization in Trouble* (W. W. Norton, 2006), http://www.earth-policy.org/Books/PB2/PB2ch10_ss4.htm.

4. Smart Growth Network and International City/County Management Association, *Getting to Smart Growth: 100 Policies for Implementation,* (January 2002), p. 10; U.S. EPA, Development, Community, and Environment Division, "Our Built and Natural Environments: A Technical Review of the Interactions Between Land Use, Transportation, and Environmental Quality" (January 2001), p. 66.

5. Bina Venkataraman, "More U.S. Commuters Drive Solo," *The Christian Science Monitor,* June 25, 2007.

6. Brown, *Plan B 2.0,* http://www.earth-policy.org/Books/PB2/PB2ch11_ss3.htm.

7. Sayeeda Warsi, "Where the Car Is Not King," BBC *Newsnight,* August 16, 2006, http://news.bbc.co.uk/2/hi/programmes/newsnight/4777801.stm.

8. Jonas Rabinovitch and John Hoehn, "A Sustainable Urban Transportation System: The 'Surface Metro' in Curitiba, Brazil" (Wisc.: EPAT/MUCIA, Madison, 1995), p. 4.

9. Lester R. Brown, *Eco-Economy: Building an Economy for the Earth* (New York: W. W. Norton and Co., 2001), http://www.earth-policy.org/Books/Eco/EEch9_ss6.htm.

10. Rabinovitch and Hoehn, "A Sustainable Urban Transportation System," p. 4.

11. U.S. EPA, "Transportation Greenhouse Gas Emissions 1990–2003," p. 10.

12. David L. Greene and Andreas Schafer, "Reducing Greenhouse Gas Emissions from U.S. Transportation," (Pew Center on Global Climate Change, Arlington, Va., May 2003) p. 10.

13. Ibid., p. iii.

14. The Electric Auto Association, http://www.eaaev.org/Flyers/index.html.

15. As calculated by Bradley Berman of Hybridcars.com, September 2007.

16. U.S. Department of Transportation Federal Highway Administration, http://ops.fhwa.dot.gov/resources/didyouknow/didyouknow_archive.asp.

17. Plug-in Partners, "Environmental Benefits of Flexible Fuel Plug-ins," http://www.pluginpartners.com/plugInHybrids/environmentalBenefits.cfm.

18. Jenna Watson, "Gas, Electric and Hybrid Go Head to Head in LCA," TreeHugger, December 23, 2006, http://www.treehugger.com/files/2006/12/gas_electric_an.php

19. Brown, *Plan B 2.0*, http://www.earth-policy.org/Books/PB2/PB2ch10_ss4.htm.

20. Greene and Schafer, "Reducing Greenhouse Gas Emissions from U.S. Transportation," p. 6.

21. Pete Harrison, "Green Appeal Helps U.K. Trains in Battle vs. Planes" (Reuters, July 3, 2007), http://www.reuters.com/article/environmentNews/idUSL2852400120070704.

7. WEEK FIVE—GREENING YOUR HOME

1. Pew Center on Global Climate Change, "US CO_2 Emissions from the Electric Power Sector," http://www.pewclimate.org/global-warming-basics/facts_and_figures/us_emissions/usco2elecpower.cfm.

2. Pew Center on Global Climate Change, "Coal and Climate Change Facts," http://www.pewclimate.org/global-warming-basics/coalfacts.cfm.

3. Interview with Arthur H. Rosenfeld, commissioner, California Energy Commission.

4. Rosenfeld, "Sustainable Development Reducing Energy Intensity by 2% per Year" (California Energy Commission, August 19, 2003).

5. U.S. Department of Energy Efficiency and Renewable Energy, "Home

Electronics Tips," http://www1.eere.energy.gov/consumer/tips/home
_office.html.

6. Environmental Defense Fight Global Warming, "Save Money and Energy at Home," http://www.fightglobalwarming.com/page.cfm?tagID=267.

7. Energy Star, "Compact Florescent Bulbs," http://www.energystar.gov/index.cfm?c=cfls.pr_cfls.

8. Helen Suh MacIntosh, "Ask TreeHugger: Is Mercury from a Broken CFL Dangerous?" http://www.treehugger.com/files/2007/05/ask_treehugger_14.php.

9. April Smith, "Building Momentum: National Trends and Prospects for High-Performance Green Buildings" (United States Green Building Council, Washington, D.C., February 2003).

10. Energy Information Administration, "Renewable Energy Sources: A Consumer's Guide" (July 6, 2007), http://www.eia.doe.gov/neic/brochure/renew05/renewable.html.

11. Interview with Eric Guyer, CEO of Climate Energy, July 23, 2007.

12. Brown, *Plan B 2.0*, p. 42.

13. John C. Woodell, Jim Dyer, Richard Pinkham, and Scott Chaplin *Water Efficiency for Your Home: Products and Advice Which Save Water, Energy, and Money*, 3rd ed. (Snowmass, Co.: Rocky Mountain Institute, 1995), p. 3; also interview with Corey Lowe, Rocky Mountain Institute, January 9, 2008.

14. Natural Resources Defense Council, "Efficient Appliances Save Energy—and Money," August 31, 2004, http://nrdc.org/air/energy/fappl.asp; also U.S. Environmental Protection Agency Energy Star program.

15. Woodell et al., *Water Efficiency for Your Home*, p. 11.

16. Union of Concerned Scientists, "Clean Energy: Solar Water Heating," http://www.ucsusa.org/clean_energy/renewable_energy_basics/solar-water-heating.html.

17. Energy Information Administration, "Renewable Energy Sources."

18. Granger Morgan, Jay Apt, and Lester Lave, "The U.S. Electric Power Sector and Climate Change Mitigation" (Pew Center on Global Climate Change, Arlington, Va., June 2005), p. 39.

19. Solar Energy International, "Energy Facts" (June 29, 2007), http://www.solarenergy.org/resources/energyfacts.html; Morgan, Apt, and Lave, "U.S. Electric Power Sector," p. 37.

20. U.S. Environmental Protection Agency, "General Information on the Link Between Solid Waste and Greenhouse Gas Emissions" (October 2006), http://epa.gov/climatechange/wycd/waste/generalinfo.html.

21. The Union of Concerned Scientists, "Clean Energy: How Geothermal Energy Works," http://www.ucsusa.org/clean_energy/renewable_energy_basics/offmen-how-geothermal-energy-works.html.

22. Ibid.

23. Frank Barnaby and James Kemp, "Too Hot to Handle? The Future of Civil Nuclear Power" (London: Oxford Research Group, July 2007), p. 8.

24. National Association of Home Builders Public Affairs and Economics, "Housing Facts, Figures and Trends," March 2006, p. 14.

25. U.S. EPA, "Outdoor Water Use in the United States" (October 2006), p. 1.

8. WEEK SIX—DRESSING UP

1. Sustainable Cotton Project, "10 Good Reasons to Join the Cleaner Cotton Campaign," http://www.sustainablecotton.org/html/manufacturers/ten_reasons.html.

2. United Nations Environment Programme, "Mining Waste: Mountains of Altered Rock, Lakes of Gleaming Waste" (2006), p. 1.

3. Sally Pick, "2004 U.S. Organic Cotton Production and Marketing Trends (Organic Trade Association, December 2005), http://www.ota.com/2005_cotton_survey.html.

4. Hollender, et al. *Naturally Clean*, p. 59.

5. Environmental Working Group's Skin Deep Cosmetics Database, "Why This Matters," http://www.cosmeticsdatabase.com/research/whythismatters.php.

6. Ibid.

7. Jenna Watson, "More Evidence that Tampons Are a Greener Choice," May 4, 2007, http://www.treehugger.com/files/2007/05/lca_tampons_pads.php.

8. Larry Thompson, "Sunscreen, Skin Cancer, and UVA" (Healthlink Medical College of Wisconsin, July 26, 2000), http://healthlink.mcw.edu/article/964647970.html.

9. EWG's Skin Deep Database, "Sunscreen Summary—What Works and What's Safe," http://www.cosmeticsdatabase.com/special/sunscreens/summary.php.

10. Jenna Watson, "Is Viscose the Way to a Greener Future?" January 25, 2007, http://www.treehugger.com/files/2007/01/is_viscose_the.php.

9. WEEK SEVEN—GETTING TO WORK

1. Energy Information Administration, "Total Energy-Related Carbon Dioxide Emissions for Selected Manufacturing Industries, 1994," http://www.eia.doe.gov/oiaf/1605/gg00rpt/images/boxtxt2fig.jpg.

2. Smithsonian National Zoological Park, "Green Team Tips," http://nationalzoo.si.edu/Publications/GreenTeam/.

3. GreenerComputing.com, "Computers Left on at Night Cost U.S. Businesses $1.7 Billion, Says Study," June 22, 2007, http://www.lohas.com/articles/100422.html.

10. WEEK EIGHT—LIVING IT UP

1. The Green Hotel Association, "How Green Are Your Travels?" http://www.greenhotels.com/grntrav.htm.

2. Bert Bras, Jay Mathewson, Michael C. Muir, "The Cycle Assessment of Film and Digital Imaging Product System Scenarios International Conference on Life Cycle Engineering" (Georgia Institute of Technology, 2006).

acknowledgments

Thank you to Kurt Edenbach, Joan Edenbach, Pat and Ed O'Neill, Sarah and Drew Fernandez, Kaila McGreal, Nancy O'Neill, Patty O'Neill, Sarah O'Connell, and Olga Sasplugas, without whom this book would not have been possible. An enormous amount of thanks must go to Julia Cheiffetz, our fearless editor, who managed to bear with us through the many changes, requests, and questions that came up during the process of putting this book together and who championed this project from the outset. Thank you to everyone at Random House and Villard who believed in and worked on this project, including Liz Cosgrove and Jennifer Rodriguez. We are incredibly indebted to and thankful for Kate Lee, our agent, who has been an extraordinary and constant source of support, knowledge, and guidance. A giant thanks goes out to the entire TreeHugger staff, especially Lloyd Alter, Mairi Beautyman, Jasmin Malik Chua, Kara DiCamillo, Sean Fisher, Sami Grover, Eva Jacobus, Jeremy Elton Jacquot, Warren McLaren, John Laumer, Federico Slivka Lederer, Christine Lepisto, Mark Ontkush, Leonora Oppenheim, Michael Graham Richard, Jessica Root, Ken Rother, Kristin Underwood, and Jenna Watson. Thank you to Felix Wittholz of Helios Design Labs for providing wonderful illustrations under tight deadlines. We would also like to thank our advocate at Discovery Communications, Claire Alexander, and the lovely and talented Discovery Planet Green team for all of their support. Thanks to Nick Aster and Simran Sethi for all of the blood, sweat, and tears they have poured into TreeHugger in the past. And for their insight and input, we also acknowledge Treehugger's board of advisors, which includes Nick Denton, Shayne McQuade, Robert M. Perkowitz, Ross Porter, Douglas Scott, and Upendra Shardanand.

The TreeHugger team currently includes: Bonnie Alter, Lloyd Alter, Paula Alvarado, Eliza Barclay, Mairi Beautyman, Ben Boyd, Jasmin Malik Chua, Erin Courtenay, Kara DiCamillo, Collin Dunn, Sean Fisher, Jacob Gordon, Sami Grover, Marisa Harris, Graham Hill, Eva Jacobus, Jeremy Elton Jacquot, Karin Kloosterman, John Laumer, Federico Slivka Lederer, Christine Lepisto, Kenny Luna, Helen Suh MacIntosh, Timothy McGee, Warren McLaren, Kimberley D. Mok, Dominic Muren, Meaghan O'Neill, Mark Ontkush, Leonora Oppenheim, Alex Pasternack, Andrew Posner, Michael Graham Richard, Jessica Root, Kelly Rossiter, Joey Roth, Ken Rother, Celine Ruben-Salama, Petz Scholtus, Vikash Singh, Matthew Sparkes, George Spyros, Annie Stahley, April Streeter, Justin Thomas, Kristin Underwood, Rachel Wasser, and Jenna Watson.

About the Authors

Photo: © Robert Severi Photography

Immersed in pop culture and ecological issues, MEAGHAN O'NEILL is a freelance writer who joined TreeHugger in 2004. Her work has appeared in a wide variety of publications, including *The Boston Globe Magazine, I.D.,* and *Teen Vogue.* She is also author of the eight-part series "The *Slate* Green Challenge with TreeHugger," a 2007 National Magazine Award finalist. O'Neill lives in Newport, Rhode Island, with her husband, son, three cats, and the world's best dog.

Photo: © Robert Severi Photography

An icon of the contemporary eco-movement, GRAHAM HILL launched TreeHugger in July 2004. Today, it has become the most-frequented green lifestyle site on the Internet. Hill is currently Vice President of Interactive at Discovery Communications, which purchased TreeHugger in August 2007. He has been pro-filed in magazines such as *Vanity Fair* and *Time,* holds a degree in architecture from Carleton University in Ottawa, and has extensive experience in industrial and Web design. He lives in New York and Barcelona.